INTELLIGENT WEARABLE INTERFACES

BICENTENNIAL
1807
⊛WILEY
2007
BICENTENNIAL

THE WILEY BICENTENNIAL–KNOWLEDGE FOR GENERATIONS

*E*ach generation has its unique needs and aspirations. When Charles Wiley first opened his small printing shop in lower Manhattan in 1807, it was a generation of boundless potential searching for an identity. And we were there, helping to define a new American literary tradition. Over half a century later, in the midst of the Second Industrial Revolution, it was a generation focused on building the future. Once again, we were there, supplying the critical scientific, technical, and engineering knowledge that helped frame the world. Throughout the 20th Century, and into the new millennium, nations began to reach out beyond their own borders and a new international community was born. Wiley was there, expanding its operations around the world to enable a global exchange of ideas, opinions, and know-how.

For 200 years, Wiley has been an integral part of each generation's journey, enabling the flow of information and understanding necessary to meet their needs and fulfill their aspirations. Today, bold new technologies are changing the way we live and learn. Wiley will be there, providing you the must-have knowledge you need to imagine new worlds, new possibilities, and new opportunities.

Generations come and go, but you can always count on Wiley to provide you the knowledge you need, when and where you need it!

WILLIAM J. PESCE
PRESIDENT AND CHIEF EXECUTIVE OFFICER

PETER BOOTH WILEY
CHAIRMAN OF THE BOARD

INTELLIGENT WEARABLE INTERFACES

Yangsheng Xu
Wen J. Li
Ka Keung C. Lee
The Chinese University of Hong Kong

WILEY-INTERSCIENCE
A JOHN WILEY & SONS, INC., PUBLICATION

Published by John Wiley & Sons, Inc., Hoboken, New Jersey
Published simultaneously in Canada

For general information on our other products and services or for technical support, please contact our Customer Care Department within the United States at (800) 762-2974, outside the United States at (317) 572-3993 or fax (317) 572-4002.

Wiley also publishes it books in variety of electronic formats. Some content that appears in print may not be available in electronic formats. For more information about Wiley products, visit our web site at www.wiley.com.

Wiley Bicentennial Logo: Richard J. Pacifico.

Library of Congress Cataloging-in-Publication Data

Xu, Yangsheng.
  Intelligent wearable interfaces/Yangsheng Xu, Wen J. Li, Ka Keung C. Lee.   p.   cm.
  Includes index.
  ISBN 978-0-470-17927-7 (cloth)
  1.   Wearable computers.
  2.   Human-computer interaction. I. Li, Wen J., 1964-. II. Lee, Ka Keung C. III. Title.
QA76.592.X98 2008
004.16 — dc22

                                                                    2007019887

Printed in the United States of America

10   9   8   7   6   5   4   3   2   1

To our families

■ CONTENTS

LIST OF TABLES

PREFACE

When I was young, my favorite Chinese classic novel book was *Outlaws of the Marsh (Shui Hu Zhuan)*. The story was about how 108 heroes banded together on a marsh-girt mountain in what today is Shandong Province in China, became leaders of an outlaw army of thousands, and fought brave and resourceful battles against pompous, heartless tyrants. The most interesting hero, in my mind, is the "Flyer on Grass" (*Cao Shang Fei*). With his "light body skill" (*Qing Gong*), he could walk on tree leaves, grass, and on water as if flying. Years later, I really had a chance to witness such martial art being performed by well-trained *kung fu* masters. One observation I made then was that the *kung fu* master could perform vertical jumps of a height many times that of his human body with the help of the tree leaves. However, he simply could not perform the acrobatic task while the leaves were not available. Then I started to think, as scientists, would it be possible for us to design and build tools or interfaces that allow humans to accomplish extremely difficult tasks that were impossible with no such interfaces? Would it be more realistic and convenient to develop such interfaces, instead of making robots, for assisting human beings to accomplish tasks otherwise impossible?

This book is a collection of such efforts we made in the Chinese University of Hong Kong. We call this research area Intelligent Wearable Interfaces. Namely, we would like to develop interfaces with a certain level of intelligence that humans can wear for enhancing their capabilities in communications, actions, monitoring, and control.

It is the our hope that this book will be of interest to human–machine interface designers and engineers who study and develop interface systems designed for wearable applications. As mobile computing, sensing technology, and artificial intelligence become more advanced and their use more widespread, we feel that the subject area intelligent wearable interfaces will grow in importance.

I would like to acknowledge the support by grants from the Research Grant Council of the Hong Kong Special Administration Region (Project CUHK 3/98C, CUHK 4197/OOE, CUHK 4228/0IE, CUHK 43 17/02E, CUHK 4202/04E), and by the Hong Kong Innovation and Technology Fund under Grant ITS/140/01.

I owe special thanks to many colleagues, students, and friends for their contributions to this book in the form of ideas, discussions, simulation, and experimental support. Among them are Ping Zhang, Wen J. Li, Ka Keung Lee, Bufu Huang, Meng Chen, Michael Nechyba, Max Meng, Kyle Chow, Weizhong Ye, Xi Shi, Huihuan Qian, and Hang Tong. Particularly important are contributions from my students and for their patience in teaching me, and from my wife Nancy and my son Peter for their moral support during the course of study.

YANGSHENG XU

We thank the technical contributions of Prof. Philip H. W. Leong, Dr. Alan H. F. Lam, Dr. Guanglie Zhang, Mr. Yilun Luo, and Mr. Chi Chiu Tsang in the work related to MIDS and μMU technologies. Many results presented in the chapters related to MIDS and μIMU are fruits of their hard work in the past few years. We are also indebted to Daka Development Ltd (especially Mr. Pat Mah) and the Hong Kong Innovation and Technology Commission (through projects ITS-185-01 and UIM-151) for their financial support in making these research endeavors possible.

WEN J. LI

Our book reviewers have contributed their advice in making critical suggestions. We are also grateful for the effort of the editors and staff (George Telecki, Kellsee Chu, and Rachel Witmer, in particular) at Wiley-Interscience throughout the development of this book.

This book could not have happened without the support from our research and development team. The work on intelligent translation glasses presented in Chapter 3 is mainly performed by Mr. Xi Shi. Mr. Cedric K. H. Law's work laid the foundation of intelligent cap interface for wheelchair control, and the work is presented in Chapter 4. Mr. Bufu Huang, Miss Meng Chen, and Mr. Weizhong Ye developed the prototypes of the intelligent shoe, which form the contents in Chapter 5. We also wish to thank Mr. Xiaoning Meng and Mr. Huihuan Qian for their editorial assistance.

KA KEUNG LEE

The Chinese University of Hong Kong
February 2007

CHAPTER 1

Introduction

Nowadays, personal communication devices such as mobile phones are very popular all over the world. The success of mobile phones lies in the fact that interpersonal communication and interface is vital to the well-being of humans, and mobile phones help people achieve this goal by extending the sense of hearing and power of speech to far away places conveniently. However, apart from the auditory sense, humans also possess other senses for communication and interaction such as vision and body motion. These sensations play very important roles in our everyday interpersonal communication. In terms of technology, the state-of-the-art in sensing hardware, control software, telecommunication protocol, and computer networks has already allowed various forms of data to be transmitted and analyzed efficiently. So why can't we find multimodal sensation communication commonplace among the mobile devices? We believe that the lack of efficient human–machine interfaces has caused a bottleneck. Therefore, we have designed and built a series of mobile devices that possess perceptual powers and communication capabilities and can support intelligent interactions. We call this novel class of devices intelligent wearable interfaces.

1.1 THE INTELLIGENT WEARABLE INTERFACE

The human interface facilitates a new form of human–machine interaction comprising a small body-worn intelligent machine. The interface is always ready, always accessible, and always with the user. It is designed to be useful in mobile settings. A wearable interface is similar to a wearable computer in the sense that they are both inextricably intertwined with the wearer, but the wearable interface does not need to have the full functionality of a computer system.

One major benefit provided by wearable intelligent interfaces is that they are in close proximity to the users so that human data such as motion and physiological information can be obtained and analyzed anywhere at anytime. The ongoing miniaturization revolution in electronics, sensor, and battery technologies, driven largely by the cell phone and handheld device markets, has made possible implementations of small wearable interfaces. Along with these hardware advances, progress in human data modeling and machine learning algorithms have also made possible the analysis and interpretation of complex, multichannel sensor data. Our intelligent wearable interface systems will leverage progress in these areas.

In this book, we present approaches to assist humans intelligently by machine learning, which are capable of being applied in pervasive and portable intelligent systems. The goal of the research is to equip interfaces and intelligent devices with the ability to detect, recognize, track, interpret, and organize different types of human behaviors and environmental situations intelligently by learning from demonstration. In terms of the interaction between humans and machines, we hope to move the place and style of human interaction away from the traditional desktops, and to put them into the real world where we normally live and act. The technologies developed in this research will allow us to enhance the effectiveness in building intelligent interfaces that are in close proximity to humans.

Instead of solving these problems using heuristic rules, we propose to approach them by learning from demonstration examples. The complexities of many human interface issues make them very difficult to be handled analytically or heuristically. Our experiences in human function modeling [1, 2] indicate that learning human actions from observation is a more viable and efficient method in modeling the details and variations involved with human behaviors. The solutions to these technical problems will create the core modules whose flexible combination will form the basis of systems that can fit into different application areas.

1.2 LEARNING FROM DEMONSTRATION

Technology and interface design issues associated with wearable computers and augmented reality displays have attracted the attention of many researchers. The previous focus was more related to ubiquitous computing and technology that attempts to blend or fuse computer-generated information with human sensations of the natural world [3], especially through personal imaging systems [4]. In this book, we will concentrate on sensor-based wearable tools that point toward practical systems that are usable by common people, not just by technical users, for applications in everyday life. As a sensory prosthesis, the main technologies of many of our application systems are based on motion sensing, wireless communication, and intelligent control with the following three major foci:

- **Hardware Design and Sensor Integration**: In designing our intelligent wearable systems, we require them to be convenient to wear and socially acceptable. Thus, the communication, sensor, and computational hardware required should not substantially change the weight and weight balance of typical clothing, lest it alter how an individual normally walks or behaves. We, therefore, in most cases, anticipate embedding the system inside human clothing, such as integrating the system into the insole of a typical shoe, spectacles, waist belt, finger-rings, or cap. These systems involve the use of vision sensors, bioelectric sensors, and motion sensors.

- **Motion Detection**: MEMS (Micro ElectroMechanical System) sensors play a major role in our endeavor to develop functional wearable interfaces because of their low cost and miniatured size. In a several chapters in our book, we present systems that use MEMS sensors to measure multi-dimensional force/acceleration of various body parts, and wirelessly transmit these motion data to the computer for information analysis. Our typical prototype consists of four main subsystems: 1) the wearable MEMS motion sensor, 2) the wearable wireless transmission board, 3) the wireless transmission interface board for PC, and 4) the information processing algorithm and display program.

- **Motion Modeling And Recognition**: Human motion is a complicated phenomenon. Meaningful information expressed by the human body is embedded in both the static and the temporal characteristics of the motions. Motion modeling is the basics of human motion analysis. It involves selection of representation form, kinetic modeling, temporal modeling, and so on. Different types of human motion have different characteristics. The adoption of recognition techniques depends on the level of abstraction where the information is to be processed. Understanding the properties of the motions to be analyzed is crucial to the design of recognition methodology.

We believe that by combining the advent in MEMS sensing, intelligent algorithms, and communication technologies, it is possible to develop novel human interface systems that could enable multifunctional control and input tasks and allow the overall shrinkage in size of the interface devices. Experimental results from our prototype systems indicate that human interfacing functions could be performed using existing vision sensors, bioelectric sensors, and MEMS-based motion sensors.

The human interface can be considered as intelligent if it employs some kind of intelligent technique to provide efficient interactivity between the user and the machine, or between the user and the environment. Example capabilities include functions such as user adaptivity, anticipation of user needs, taking initiative and making suggestions to the user, and providing explanation of its actions. We believe that this research area will open up tremendous new human–computer interface possibilities, resulting in rich academic research contents and potential product lines in consumer electronics and multimedia industries.

This book is composed of eight chapters. This first chapter serves as an introduction and overview of the work, and related research is summarized. In Chapter 2, we present the network architecture system for a wearable robot, which is a mobile information device capable of supporting remote communication and intelligent interaction between networked entities. Chapter 3 addresses the intelligent glasses, which can automatically perform language translation in real time. In Chapter 4, we introduce the intelligent cap that enables a severely disabled person to guide his/her wheelchair by eye-gaze. In Chapter 5, we develop a sensor-integrated shoe system as an information acquisition platform to sense foot motion. It can be used for computer control in the a form of a shoe-mouse, human identification under the framework of capturing and analyzing dynamic human gait, and health monitoring for patients with musculoskeletal and neurologic disorders. In Chapter 6, we present our work on merging MEMS force sensors and wireless technology to develop a novel multifunctional finger-ring–based interface input system, which could potentially replace the mouse, pen, and keyboard as input devices to the computer. In Chapter 7, we present the motion-sensor–based Digital Writing Instrument for recording handwriting on any surface. In Chapter 8, a human airbag system is designed to reduce the impact force from slippage falling-down. A recognition algorithm is developed for real-time falling determination.

REFERENCES

1. M. C. Nechyba and Y. Xu, "Human Control Strategy: Abstraction, Verification and Replication," *IEEE Control Systems*, Vol. 17, No. 5, pp. 48–61, 1997.
2. K. K. Lee and Y. Xu, "Computational Intelligence for Modeling Human Sensations in Virtual Environments," *Journal of Advanced Computational Intelligence*, Vol. 8, No. 3, pp. 302–312, 2004.
3. W. Barfield and T. Caudell, Eds., *Fundemantals of Wearable Computers and Augmented Reality*, New Jersey, Lawrence Erlbaum Associates Publishers, 2001.
4. S. Mann, *Intelligent Image Processing*, New York, IEEE and Wiley, 2002.

CHAPTER 2

Network Architecture for Wearable Robots[1]

2.1 INTRODUCTION

We have developed a novel robot concept called the wearable robot, and we will present the development of the supporting network architecture. Wearable robots are mobile information devices capable of supporting remote communication and intelligent interaction between networked entities. In this chapter, we explore the possible functions of such a robotic network and will present a distributed network architecture based on service components. To support the interaction and communication between the components in the wearable robot system, we have developed an intelligent network architecture. This service-based architecture involves three major mechanisms. The first mechanism involves the use of a task coordinator service such that the execution of the services can be managed using a priority queue. The second mechanism enables the system to automatically push the required service proxy to the client intelligently based on certain system-related conditions. In the third mechanism, we allow the system to automatically deliver services based on contextual information. Using a fuzzy-logic-based decision-making system, the matching service can determine whether the service should be automatically delivered using the information provided by the service, client, lookup service, and context sensors. An application scenario has been implemented to demonstrate the feasibility of this distributed service-based robot architecture. The architecture is implemented as extensions to the Jini network model.

Nowadays, robots can perform tasks in many different places, environments, and settings. For example, we can find field robots working in dangerous areas, service robots employed in households, medical robots performing operations in hospitals, industrial robots working in factories, and so on. To further explore the potential impact of robotic technologies on our quality of life, we may consider a new

[1]Reprinted, by permission, from Ka Keung Lee, Ping Zhang, Yangsheng Xu, and Bin Liang "An Intelligent Service-Based Network Architecture for Wearable Robots," *IEEE Transactions on Systems, Man and Cybernetics, Part B: Cybernetics*, August 2004, Volume 34, Number 4. Copyright © 2004 by IEEE.

robotic niche where robots are wearable, in close proximity to humans, and can accompany us wherever we go. We are exploring the concept of wearable robots that can be conveniently carried by humans, be worn on human bodies, and are capable of supporting various kinds of communication and interaction with the users. Wearable robots are interfaces through which users can interact with other people and other networked machines. We view wearable robots as mobile information devices that are capable of supporting remote communication and intelligent interactions between networked entities. They help users to extend their sensing and actuating capabilities through a networking environment, as well as receiving access to useful information.

We have come some way in establishing the methodologies required for developing distributed wearable robot systems, which can be used to support mobile devices for communication and interaction. Our research will provide a framework for constructing distributed robot infrastructures in which each robot is a portable communication interface through which remote interaction among humans, robots, and machines in the networked environment can be achieved. In this chapter, we will address two main issues: 1) how to develop a distributed robot architecture that can efficiently support the communication and control among robots, humans, and other machines in the networked environment, and 2) how to incorporate contextual information processing in the wearable robot architecture.

Currently, typical communication technologies, such as mobile phones, e-mail messages, and video-conferencing, can only offer limited means for information conveyance. Most of these mobile communication devices lack the abilities to support intelligent interaction, not to mention more complicated functions such as visual perception and motion feedback. By using a wearable robot as a communication interface, we can explore more possibilities in human–human and human–machine interactions that are not yet common. In this research, the notion of robot actuators are not only confined to manipulators that handle physical objects nearby, but also they include any physical medium that conveys information to the user. These functional units, such as an LCD display or a small loudspeaker, can be considered as actuators that influence perceptions.

Substantial research has been carried out in relation to human interaction with intelligent robots. Much has concentrated on considering the human–robot interface as part of the robot system [1], but our work points out that the robot itself can be an interface to other people and machines. Special problems, such as sensor stability and adaptability of interaction, arise when the interface is engaged in different mobility modes together with the user. Preliminary research in association with interface adaptability has begun in the application area of mobile computing devices [2], and we think that these advancements can be fused together with robotic technology and distributed network communication to form the foundation of the wearable robot system.

The rest of the chapter is organized as follows. In Section 2.2, we will discuss the types of interaction made feasible by using our intelligent distributed wearable robot architecture. In Section 2.3, the design of a prototype of the wearable robot will be presented. The detailed design of the service-based network architecture of the wearable robot system will be introduced in Section 2.4, including three kinds of extension to the original Jini network model, along with the functions of the matching service and task coordinator service. We have developed an application to demonstrate the feasibility of the wearable robot system, which is explained in Section 2.5. Our work will be compared with other previous works in Section 2.6. Section 2.7 contains the conclusion of this work.

2.2 WEARABLE ROBOTS AND INTERACTIONS

We are developing a robot system that can support interaction among different entities (Figure 2.1). Assume that robot 1 and its human owner 1 are situated in environment 1, and that robot 2 and its human owner 2 are situated in environment 2. This simple scenario may give rise to different types of possible interactions. First of all, the robot can interact with its owner through its intelligent interface. The robot can be equipped with different sensors to detect its owner's intentions. The robot can also express its operational state and feedback using its actuators (A). For example, the user may issue a command by producing certain facial gestures to express his/her intention, which the robot will recognize by using its vision system. The robot can then reflect its understanding by nodding for "yes," turning its head for "no," or a hand motion for "I don't understand." The interaction between the human and the environment can be observed and analyzed by the robot using various sensors such as motion and vision sensors (B). The robot can also measure the current state of the environment using embodied sensing capabilities (C) and act to influence the environment by controlling networked actuators (D). For example, using its temperature sensor, the robot may determine that the office temperature is not suitable for its owner. Having obtained information from the room's security system that reveals its master is the only person in the room, the robot may then issue a command to the

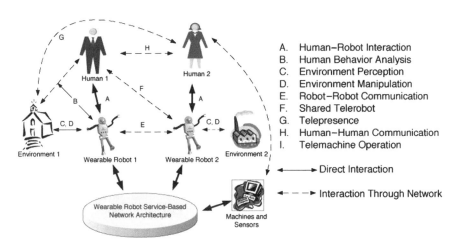

Figure 2.1. Types of interactions supported by the system.

temperature control system of the office to have the room temperature adjusted. In addition, the communication between different robots allows them to exchange information and interact with each other (E). A human can take control over another user's robot and pass the control of his own robot to another robot user (F). This process may proceed after the direct negotiation between the users or upon the satisfaction of certain agreed conditions. In the former case, the robots act as interfaces for human–human communication (H). Moreover, under mutually accepted terms, a user may explore the environment where another remote robot is situated (G), such as using another user's robot as a remote sensor. Finally, robots can be used to control other networked machines or be used to reveal the state of those machines (I), such as sending an image taken by the robot directly to the nearest printer for printing.

Apart from the capability of supporting the interaction among robots, humans, and networked machines, contextual information processing is also a key technology incorporated into our distributed service architecture. By making the applications context-aware, we can increase the amount of situational information made available to robots. Recent development in the area of context-awareness computing [3–5] has helped us to identify important types of context information, which may include 1) system environment, such as network capability, connectivity, and cost of computation; 2) user environment, such as location, identity of nearby people, and social situation; and 3) physical environment, such as lighting, temperature, and noise level.

2.3   WEARABLE ROBOT DESIGN

We have developed a wearable robot prototype by integrating various micro-sensors, actuators, and learning technologies (Figures 2.2 and 2.3). A wearable robot is characterized by properties desirable for mobile devices, such as small in size, light in weight, and flexible in functionality. It is equipped with a mini-camera for

Figure 2.2. Prototype of wearable robot.

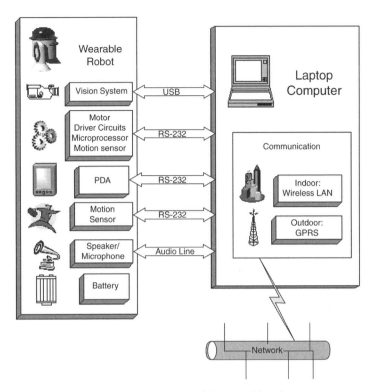

Figure 2.3. Hardware architecture of the wearable robot prototype.

visual information gathering, a motion sensor for analyzing robot and user movements, a touch sensor for interacting with users based on contacts, and so on. In the current design, the actuators and manipulators of the robot have three main functions. They allow the robot to 1) perform various expressions for communication purposes, 2) control the position of the camera for the active vision system, and 3) be used as a pointing device for locating different objects in the surrounding environment [6]. The personal digital assistant (PDA)-based interface processor further enhances the interface capability of the robot by using communication means such as text and graphics display, video streaming, and music. The prototype version of the wearable robot has four joints, and they are actuated by micro-servomotors. The motors control the pan and tilt angles of the camera and the motion of the two arms. The positions of the motors are controlled using the Proportional-Integral-Derivative (PID) scheme. The visual information and PDA communication signals are processed by a micro-computer based on PC/104 technology. The motion sensor (Crossbow CXTA02, Crossbow Technology Inc., San Jose, CA) signals and servomotor control signals are handled by a servomotor controller board equipped with an analog-to-digital conversion capability (Pontech SV203B, Pontech, Rancho Cucamonga, CA), and the motor controller board is connected to the computer board via RS-232. The computer communicates with the rest of the computer network using a wireless local area network (LAN) while indoors and a general packet radio service (GPRS) when outside.

To help the wearable robot to understand the intention of its user, it should be aware of the affect of humans [7]. Since the wearable robot is in close proximity to humans while in operation and the human face is a natural medium for intimate human expression, we have developed a visual learning system that allows the wearable robot to identify the intensity of emotional expression on the user's face [8]. Using a statistical learning technique called support vector regression, the vision system on the wearable robot can track the face of its user and estimate the intensity of facial expressions for emotions such as happiness, anger, and sadness. As shown in Figure 2.2, our wearable robot prototype can be worn conveniently around the neck, whereas the PC/104-based computer is carried in a small bag around the waist. However, for certain applications, such as emotion recognition and text-based interaction, the robot will be held in the user's palm like any handheld device. We envision that wearable robots are suitable for at least the following types of applications: 1) applications that can benefit from physical interactions between the robot and user, such as manipulator gesture motion, force feedback, and limited manipulation of small objects; 2) applications that require motion control of actuators and sensors, such as visual servoing, active proximity sensing, or visual–vestibular integration for stabilization; and 3) applications that require portable devices to be used in different levels of proximity to the user. For example, the wearable robot can be worn on the body, carried by hands, or placed on a desk for interaction.

2.4   DISTRIBUTED SERVICE-BASED ARCHITECTURE

In terms of robot network architecture, different approaches have been proposed by researchers [9–11], covering the deliberate planning, the reactive behavioral, and the hybrid methods. They have all proven successful in managing autonomous robots and multi-robot systems to varying degrees. Many distributed robot and agent systems are built around a message-based architecture in which the system components are tightly coupled and include strict data standards for information exchange [12]. In the traditional client–server model, a client typically interacts with a specific server, usually one with a known interface and known location, which means clients and servers are tightly coupled. However, as the number and type of networked members increase, it is not sufficient to rely on a system that can only provide an uncoordinated stateless exchange of information. To support the dynamic reconfiguration of components in the network system, the interoperability of components cannot be defined by the data standard alone, but they should also be defined by the functional standard. That means that components in the system should be identified and accessed according to the functions they can provide.

We have thus developed an intelligent service-based architecture to support communication within a distributed wearable robot system. Service-based architecture breaks the tight coupling between components by providing a higher abstraction layer called service, and it has attracted much attention in recent years. Similar architectural concepts can be found in applications, including telemedicine [13], Web application rapid prototyping systems [14], agent-based telecommunication systems [15], and distributed symbolic computation systems [16]. Using a service-based architecture, our distributed robot system enjoys several advantages offered by object technology [17]. First, it permits the creation of service in such a way that its user may be another service and not necessarily humans. The resources used by a service can also be provided by another service, which makes complex interactions between robots and other networked machines possible. Second, it allows service re-usability by permitting the specification of new services relying on existing specifications. The facilities provided by existing services can be used to create new services. Third, it can to support integrated design of management and control by encapsulating, in distinct interfaces of the same object, operations related to specific jobs and those dealing with management of that object.

In a service-based architecture, typical interaction usually occurs between the service and the client. The service in essence publishes itself by providing a general, high-level description of the location, category, and function it provides. It also reveals the technical details of its network location through which it can be connected. On the client side, it needs to determine the service required and make relevant enquiries to the network. After obtaining a reference to the service implementation, the client can invoke the service.

In our implementation, we apply a Jini-based architecture to support the distributed communication. Jini [18] is a middleware whose purpose is to federate groups of devices and software components into a single, dynamic distributed system to provide simplicity of access, ease of administration, and support for sharing. A basic Jini system is composed of three main parts. First, there is a set of components that provide an infrastructure for federating services in a distributed system. Second, Jini specifies a programming model that supports the production of reliable distributed services. Third, there are services that can be made part of a federated Jini system and can offer functionality to members of the federation.

In our wearable robot system, the wearable robot functions are implemented as services, which are defined as entities that can be used by a person, a program, or other services. Members of the Jini system federate to share access to robots. Therefore, it is more than just a set of clients and servers. Robots in the system can make use of other services, and a client of one service may itself be a service with clients of it own. Using this architecture, a service or robot can be plugged into the network and is visible and available to those who want to use it. Users do not need to be aware of which part of the services they want to use that is implemented in software or hardware. When services and robots are plugged into the network and are available, they can be used by clients and other services directly. A lookup service maps interfaces indicating the functionality provided by a service to sets of objects that implement the service. A robot or a service can be added to a lookup service by a pair or protocol called Discovery/Join.

The service-based architecture using Jini can provide several advantages to the wearable robot system. First, we want to develop a network system in which the services, including robot, machines, and other functional devices, can find each other automatically. Jini helps by providing a lookup service that enables any networked member to identify and gain access to a service by specifying the desired functions. Second, it is desirable that components can be added to and removed from the network without affecting other components. During this operation, communication relationships between services and components may change. The components in the service-based architecture are loosely coupled and network stability can be improved by Jini by ensuring such a change is carried out seamlessly. Third, wearable robots are portable and their locations may change as the users move. Wearable robots can be service providers or clients. As service receivers, services delivered to the wearable robots may need to change destination as the robots move. As service providers, the positions from where the robots provide their services may also change as they move. By using service stub mobility, the Jini system allows mobile services to be accessed by the clients in the network.

2.4.1 Extension to the Jini Model

The main components in the original Jini network infrastructure are lookup services, proxy objects, and the Discovery/Join protocol. The lookup service is the most important service in the Jini federation, and it maintains a repository of service providers. Before the services can be used by other components in the system, they must register with the lookup service (Figure 2.4 and Table 2.1). The lookup service then acts as an interface through which services can be located, thus enabling users to interact with services. Through the Discovery/Join protocol, new services can be added to the federation. The lookup protocol then allows the members of the network to locate and access the services provided by other members. If a new service would like to advertise its presence, it can drop a multicast packet by using the multicast request protocol to a defined port. As the lookup service listens to the port, it has the capacity to pick up the discovery request packet and retrieve the contact address of the new service so that they can establish a communication link using the transmission control protocol. If a client wants to obtain access to a service in the federation, it can send a request to the lookup service specifying the required service details. If the service is available, then the proxy object of the service provider will be downloaded to the Jini client so that the Jini client can communicate with the service provider directly through the proxy.

The original service delivery model is a pull model in which the clients request the services and the service proxies are delivered to them if the services are available. To

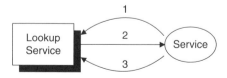

Figure 2.4. Service registration.

Table 2.1. Steps for Service Registration

Step	From	To	Action
1	Service	Lookup Service	Send multicast request packet
2	Lookup Service	Service	Receive information from lookup service
3	Service	Lookup Service	Transmit proxy to lookup service

handle the issues related to efficient resources allocation and context-aware applications in our wearable robot system, we develop three types of extension to the
original Jini network architecture. The extensions involve the introduction of two
new services—the matching service and the task coordinator service. The first extension is an enhancement of the pull model, and the second and third extensions are
based on the service push model.

The matching service decides whether a service should be provided to a particular
client, based on a fuzzy decision-making process influenced by the conditions of the
service provider, the client, the network system, and the environmental context information. The function of the fuzzy-based matching service will be further discussed in
Section 4.2. When the system starts up, the services will form a federation by regis-
tering their presence by transmitting their proxies to the lookup service. Any new
service can join the federation dynamically after the lookup service has started up.
Clients who wish to utilize the services in the federation will need to contact the
lookup service (Figure 2.5).

In the first mechanism, we add a task coordinator to the original Jini structure such
that task priority can be taken into account. This is a service pull model in which the

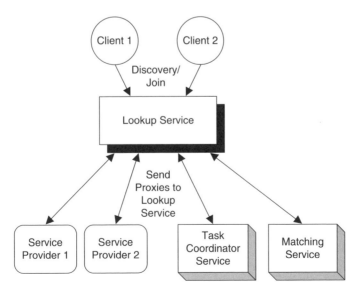

Figure 2.5. All clients and services must contact the lookup service upon joining the
federation.

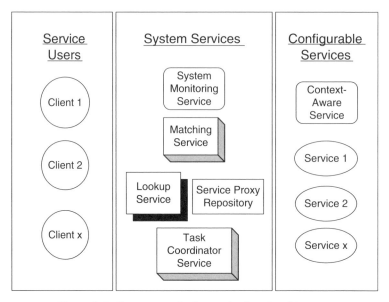

Figure 2.6. Components in the service-based architecture.

required services can be pulled by the human users or automatically pulled by the clients such as the wearable robots. The second and third mechanisms support the service push model. In the second mechanism, service proxies are pushed to the clients automatically based on the information provided by the matching service. In the third mechanism, the service proxies are not pushed to particular clients, but the services are automatically delivered based on the results of the decision-making process of the matching service.

There are three types of networked entities in our design of the service-based architecture. They are the system services, the configurable services, and the service users (clients). The system services include the lookup service, the system monitoring service, the matching service, and the task coordinator service. The task coordinator service contains a queuing system that manages the execution of the service tasks. Using fuzzy logic, the matching service determines the priority levels of the service tasks to be executed. The system monitoring service mainly keeps track of the number of running services in the system and the number of queued tasks. For context-aware applications, context-aware services are used to manage the context related sensor data. We will explain the extensions below.

2.4.1.1 Extension 1: The Addition of Task Coordinator Service. The limited resources in distributed systems affect the provision of services to clients from time to time. The task coordinator service provides coordination between all tasks by considering their content and priority. To meet the requirement of the quality of service in the system, a queuing mechanism is implemented in the task coordinator such that the resources in the system can be controlled. After a client

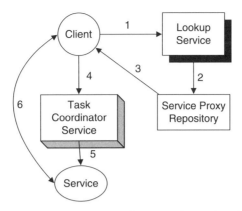

Figure 2.7. Incorporation of the task coordinator service to the original Jini model.

has downloaded the proxy of a service provider from the service proxy repository, it has to notify the task coordinator before the service can be delivered. The steps for doing so are shown in Figure 2.7 and Table 2.2.

The task coordinator service contains a first-in–first-out queuing system, which allows task reordering by priority. It also coordinates the execution order of the services using priority queues. Services having equal priority levels are executed in a first-in–first-out manner. The task having the highest priority in the sequence is executed first (descending order), and there is no assumption on how elements enter the priority queue but only a criterion for their exit. There is one queue for each service provided by each server, and the queues are implemented using a heap data structure. The services will notify the task coordinator service when they are available for delivering new services. The priority levels of the services are determined by the matching service using fuzzy logic.

Table 2.2. Steps Involving the Task Coordinator Service

Step	From	To	Action
1	Client	Lookup Service	Client discovers and joins the lookup service
2	Lookup Service	Service Proxy Repository	Request service from service proxy repository
3	Service Proxy Repository	Client	Proxy downloaded from Web server
4	Client	Task Coordinator Service	Send task object to task coordinator, join the queue, and wait for service task execution
5	Task Coordinator	Service	When the service is ready for use, the task of the client will be sent to the service
6	Client	Service	Service is delivered to client

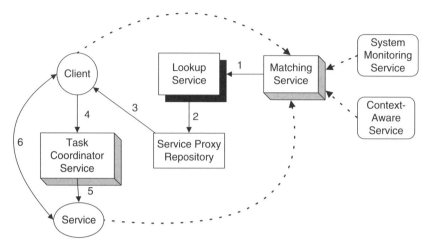

Figure 2.8. Pushing the service proxy automatically to the client based on the processing result of the matching service.

2.4.1.2 Extension 2: Automatic Service Proxy Download.

If the system is to deliver services efficiently in applications, not only should users have the ability to pull the required service for utilization, but also more importantly, the system should be able to intelligently push the services to the users when certain conditions are met. In this extension, the system is equipped with a decision-making service, called the matching service, which monitors the relationship among the service provider, the client, the network system, and the environmental context. The aim is that when the matching service has reached the conclusion that the collective behavior of all parties in the system has come to a certain state, the proxy of the suitable service will be pushed to the clients who need them. The steps for this process are shown in Figure 2.8 and Table 2.3.

Table 2.3. Steps for Automatic Service Proxy Push

Step	From	To	Action
1	Matching Service	Lookup Service	Lookup service receives information from the matching service
2	Lookup Service	Service Proxy Repository	Request service from SPR
3	Service Proxy Repository	Client	Proxy pushed to client
4	Client	Task Coordinator	Send task object to task coordinator, join the queue, and wait for service task execution
5	Task Coordinator Service	Service	When the service is ready for use, the task of the client will be sent to the service
6	Client	Service	Service is delivered to client

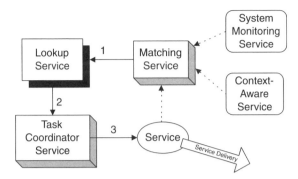

Figure 2.9. Automatic delivery of service based on the processing results of the matching service.

This procedure provides the system with the intelligence needed to avoid downloading all the unnecessary codes for which the robot may not have enough memory to handle. This arrangement also reduces the load on the user by hiding the decision-making process in the background.

2.4.1.3 Extension 3: Automatic Service Delivery. Our third extension to the Jini architecture is also based on a push service model. For certain services, there may be particular clients to which the proxies are not required to be delivered. Instead, the services can be directly performed without establishing links between the client components and the service providers. This occurs when the client of the service is a human but not a network component. The automatic delivery of the services is determined by the condition of the service provider, the system status, and the contextual data obtained from sensors. In this case, the matching service does not require input from a client component. The steps for this process are shown in Figure 2.9 and Table 2.4.

2.4.2 The Matching Service

We apply fuzzy logic to analyze the collective behavior of the services provided by wearable robots and the components in the network system. Fuzzy logic [19] is an

Table 2.4. Steps for Automatic Service Delivery

Step	From	To	Action
1	Matching Service	Lookup Service	Lookup service receives information from the matching service
2	Lookup Service	Task Coordinator Service	Service task is pushed to task coordinator service
3	Task Coordinator Service	Service	When the service is ready for use, the service will be automatically delivered

approximate reasoning technique based on linguistic terms, and it has been applied in different areas in network communication [20] such as server selection [21], queuing system [22], and load balancing [23]. Fuzzy logic is a suitable technique for application in the matching service for making decisions relating to service delivery for a number of reasons. First, an accurate state of the global network is difficult to update and maintain. In a dynamic network, the global state will always be imprecise and this imprecision will only increase as more interacting components are added to the federation. Second, good and realistic quantization can be provided by fuzzy states, since the system and human related features only have limited resolution. Unrealistic results may occur if computation is made using single values or sharp intervals. Third, our network takes context information into account when making the decisions related to automatic service delivery. Context information is usually described using linguistic terms. For example, when we express the context of the relative position between two objects, we often describe them as being "close," "very close," or "not too close," which are fuzzy.

The purpose of the matching service is that after observing the interaction between the clients and the service provider, it can abstract the relationship between the system context and the delivery of services, such that when the appropriate conditions are met, the suitable services will be automatically delivered to the clients.

Four different indices are fed into the matching service for the purposes of decisions related to service push, which include the service supply index, the

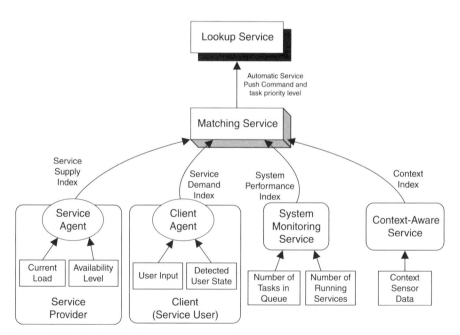

Figure 2.10. Relationship between the matching service and other components in the federation.

service demand index, the system performance index, and the context index. The matching service is a decision-making engine based on fuzzy logic. The context index is calculated by the context-aware service based on the context sensor data. The context-aware service provides the functions of data abstraction and separation of concern. For simple cases, the context-aware service can perform a straightforward scaling operation on the sensor data for normalization. For more complex context types, additional sophisticated mechanisms such as those provided in the Context Toolkit [3] can be used to help handle the process. The service supply index, the service demand index, and the system performance index are generated by fuzzy engines based on two inputs each. The system monitoring service observes the number of services running in the system and monitors the number of tasks in the queue inside the task coordinator service. The service demand index is generated by a client agent residing in the client. The client agent keeps track of the user's state (by using intelligent user interfaces) and receives direct input from the user indicating the level of desire for a particular service. The service supply index is also generated by a service agent residing within the service provider. By detecting the relationship between the current load of the service provider and how much the service would like to make itself available to the clients, the service agent determines the service supply index of this service. When the matching service receives a new non-low index from the client agent or the context-aware service, the matching service will request the indices from the service agent and the system monitoring agent. The decision-making process will start after the three required indexes are all obtained, two of which must be the system and service supply indexes.

Fuzzy decision systems are used in service agents, client agents, the system monitoring service, and the matching service. The implementation of the fuzzy system includes the following steps: 1) fuzzification of inputs, 2) application of fuzzy inference rules, and 3) defuzzification of the output. Linguistic rules designed based on human experience are applied in the inference process. The membership function chosen for the input fuzzy set is triangular, which is due to its simplicity and efficiency in description and storage. It can also be used to approximate other membership functions [24]. The linguistic terms we use in the fuzzy system include high, middle, and low. The universe of discourse for the fuzzy inputs is [0,4] because it is convenient and frequently used for domains with three fuzzy sets. The membership function for the output fuzzy set of the matching service is rectangular, and the universe of discourse for the fuzzy output is [0,3]. The AND operation is implemented as *min*, and the OR operation is implemented as *max*. The rule base of the inference systems is shown in Tables 2.5 to 2.8.

The *max* operation is used for the aggregation of different rules. The defuzzification step in the service agent, client, and the system monitoring agent is achieved by using the center of gravity operation. For environmental context-sensing applications, the role of the service demand index will be replaced by the context index. The context index signifies the degree of presence of particular combinations of context elements that are of interest to the application. The way in which the context index is generated by the context service is sensor and application dependent.

Table 2.5. Rule Base for Determining the Service Supply Index

Load	Availability	Service Supply Index
Low	Low	Low
Low	Middle	Middle
Low	High	High
Middle	Low	Low
Middle	Middle	Middle
Middle	High	Middle
High	Low	Low
High	Middle	Low
High	High	Low

Table 2.6. Rule Base for Determining the Service Demand Index

User Input	Detected User State	Service Demand Index
Low	Low	Low
Low	Middle	Middle
Low	High	High
Middle	Low	Middle
Middle	Middle	Middle
Middle	High	High
High	Low	High
High	Middle	High
High	High	High

Table 2.7. Rule Base for Determining the System Index

Number of Tasks in Queue	Number of Services Running	System Index
Low	Low	High
Low	Middle	High
Low	High	Middle
Middle	Low	Middle
Middle	Middle	Middle
Middle	High	Middle
High	Low	Low
High	Middle	Low
High	High	Low

Table 2.8. Rule Base for Matching Service

Service Supply Index	Client Demand Index or Context Index	System Index	Action
High	High	High	Auto Proxy Push or Service Delivery
High	High	Middle	Auto Proxy Push or Service Delivery
Middle	High	High	Prompt User
Middle	High	Middle	Prompt User
High	Middle	High	Auto Proxy Push or Service Delivery
High	Middle	Middle	Prompt User
Middle	Middle	High	Prompt User
Middle	Middle	Middle	Prompt User

If (Service Supply Index = Low) OR (Service Demand Index = Low) OR (Service Demand Index = Low), then Action = No Action.

The basic rationale behind the fuzzy rule base is as follows:

1. The service supply index will be higher if the service has a higher availability or a lower load.
2. The service demand index will be higher if the client has a higher detected need or a higher input desire to use the service. The needs of the user can be detected by an intelligent user interface.
3. The system index will be higher if the number of items in the queue is lower or the number of running services is lower.
4. If enough of the input indices are sufficiently high, the service proxy will be automatically pushed to the client or the service will be automatically delivered. If the indices are too low, no action will be taken by the lookup service. If the indices are close to middle value, the user will be prompted to make a decision.

2.5 APPLICATION SCENARIO

To evaluate the performance of the proposed system, several initial experiments are carried out on a simplified setup. The HTTP server, the Java Remote Method Invocation System Daemon (RMID), lookup service, and clients all run on a Celeron PII 566 MHz machine with 256 MB RAM. We tested the registration and search times of services under different conditions. The relationship between the time to register a new service and the number of registered services in the lookup service is plotted in Figure 2.11. The relationship between the time to search for the lookup service or a normal service and the number of registered services is

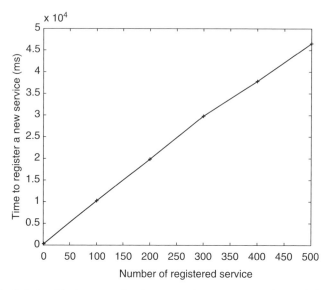

Figure 2.11. Relationship between the time to register a new service and the number of registered services.

plotted in Figure 2.12. The results show that 1) the number of services registered in the system has no significant influence on the performance of the registered services (such as the lookup service), 2) the number of services registered in the system has no significant influence on the time required to search for a service, and 3) there is a

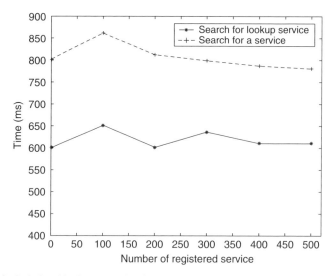

Figure 2.12. Relationship between the time to search for the lookup service or a normal service and the number of registered services.

linear relationship between the time required to register a new service and the number of services already registered in the system. Figure 2.12, shows that the system peaks at roughly 100 services. Compared with the amplitude along the time axis, the peak is relatively small. This may be a consequence of other processes in the operating system that suddenly occupy central processing unit (CPU) resources. The results obtained here provide us with valuable information for the detailed design of applications and characterization of the architecture, such as the determination of the optimal number of services supported by each lookup service, communication and interaction timing analysis, and optimal resource allocation.

For the purpose of demonstrating the usefulness and feasibility of the service-based architecture for networked wearable robots, we have implemented an application system in our research lab. Let's consider a scenario situated in an office and a lab using the wearable robot and the service-based architecture for context-aware and communication purposes (Figure 2.13). In this application, we use a visual person-tracking system to detect the presence of a visitor in the lab and use location context information to determine which service to activate. Apart from the visual person-tracking system, the lab is equipped with two voice communication service providers (A and B) and a humanoid robot. If the visitor approaches the humanoid, it will present a greeting message to the visitor. If the visitor leaves the lab, the humanoid will present a farewell message. The master of the wearable robot, who is working in another office, will be notified by the wearable robot if the visitor moves near one of the voice communication service providers. A voice communication link can then be established between the wearable robot's master and the visitor in the other room through microphones. If abnormal behavior is displayed by the visitor, such as moving quickly or running or remaining stationary for an extended period, the wearable robot user in the office room will also be notified and he/she can request control over the motion of the humanoid robot for a period of time by use the interface on the wearable robot. The flowchart of this application scenarios is shown in Figure 2.14.

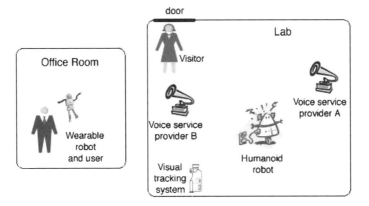

Figure 2.13. Setup of the two rooms in the application.

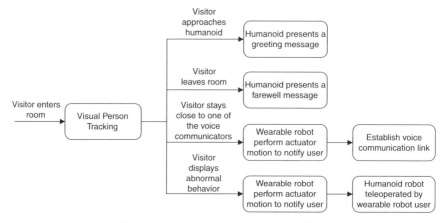

Figure 2.14. Flowchart of the application.

The visual person-tracking system, the voice communication gateway, the humanoid action controller, and the wearable robot action controller are implemented as Jini services. The person-tracking system can locate the position of a visitor by using a background subtraction technique and thus determine the normality of the walking trajectory [25]. The visual system can also detect if the visitor is displaying a high degree of motion by calculating the rate of change in the person's rectangular shape and size. The context index is directly related to the person's position. One context index is related to each specified position in the lab, including the areas close to the door, the humanoid robot, and the two voice communication service providers. For example, the position context index related to the humanoid robot would give a higher value if the person is closer to the humanoid robot.

There is one entry in the matching service for each service. The input configuration for each matching service entry is designed before service registration. The input of the matching service must include the service supply index and the system performance index. The service demand index and the context index are optional, but at least one of the two must be present. Depending on the design of each service, it can be triggered by the client or by a contextual information change in the environment. The details of the implemented services are shown in Table 2.9.

The context information service provided by the visual tracking system is a continuous sensing service and is always on. The humanoid teleoperation service allows its client to take control over the motion of the humanoid robot for a certain period of time. These two services are provided to any capable clients that request it and is not dependent on any other context information for activation. The humanoid farewell service and the humanoid greeting service will be automatically delivered if the matching service considers that the required conditions are met and that these two services do not require client information. The wearable robot user notification service allows the wearable robot to be notified by the person-tracking system. The context index of its matching service will be high if the visual tracking system finds that the visitor stays in front of the humanoid or that abnormal human behavior is

Table 2.9. Services Developed for the Application Scenario

Service Name	Service Provider	Client	Context Data
Person-tracking service	Visual tracking system	Any services using this context information	Not needed
Humanoid robot teleoperation service	Humanoid robot	Wearable robot	Not needed
Humanoid greeting service	Humanoid	Human visitor	Human approaches humanoid
Humanoid farewell service	Humanoid	Human visitor	Human leaves room
Wearable robot user notification service	Visual tracking system	Wearable robot	Human stays in front of humanoid or abnormal human behavior detected
Voice communication service	Voice communication service provider A and B	Wearable robot	Human stays close to voice communication service provider A or B

displayed by the visitor. The voice communication service allows us to experiment with the effect of the automatic service proxy push mechanism. Depending on which voice communication service provider (A or B) with which the visitor is in close proximity, the corresponding service proxy will be pushed to the wearable robot to allow both users to establish a communication link.

The application scenario has been successfully implemented using our service-based architecture. Most of the services and components are written in the Java programming language, but there are some exceptions. For example, the visual tracking system is implemented using C + + and is wrapped around by Java interfaces.

2.6 RELATED WORKS

In this section, we will compare our research against other works in three main areas: intelligent mobile devices, Jini application systems, and network architectures that support intelligent environments and robots.

In the wearable computing area [4, 26], numerous types of intelligent devices have been developed for the purposes of sensing, information transfer, data storage, and communication. For example, Tsumaki and Uchiyama [27] presented the design of a personal telerobotic system that comes in two different forms: wearable and mobile. The system can be used for communication via the Internet. However, the author did not explain how the network communication is achieved and no prototype is shown. Schmidt et al. [28] integrated an accelerometer and two microphones on a tie to collect a user's related context information. Ljungstrand et al. [29] introduced a public wearable device whose hardware is based on a modified GameBoy$^{©}$ which is mainly used for information display and requests. Kato et al. [30] developed a wearable active vision system for distributed computing environments. A camera is mounted on the user's shoulder, and a PDA-based interface is placed around the wrist. Mayol et al. [31] presented a wearable active visual sensor that can achieve a level of decoupling of camera movement from the wearer's posture and motions. This device is also mounted on the user's shoulder. Starner et al. [32] presented a wearable device capable of recognizing human hand gestures and controlling home automation systems. A small camera is worn as part of a necklace or pin.

Our wearable robot design differs from the previous works in a number of ways. 1) The previous wearable devices do not make use of actuated mechanisms for expression, whereas our wearable robot can use its limb motions to express itself. 2) The outputs and inputs of the wearable robot are both multimodal (visual recognition, motion sensing, touch sensing, voice sensing, text and graphics display, sound speaker, and actuated motion. 3) The concept of our system is more like a networked robot rather than just a typical mobile device. Our definition of a robot is that the machine possesses sensing, actuation, and an intelligent decision-making capability. 4) Our wearable robot also possesses a certain level of autonomy. When used in the autonomous mode, the user can play with the wearable robot as if the robot possesses a range of emotions. 5) Based on the current design, the relative position between the wearable robot and the human body is not fixed. For example, it can be worn around the neck for context information collection (wearable), held by the user's hand as a face-to-face communication device (portable), and can also be placed on a desk for information display and event notification (desktop device).

Our system is currently implemented in an indoor environment, and we will extend its functionalities such that it can work outdoors in the future. In the literature, there are many designs for distributed communication systems and network architectures for ubiquitous computing [33–35]. However, there are few that support the interaction among humans, robots, and wearable devices in intelligent environments. Lee et al. [36] developed an intelligent space equipped with sensors, actuators, and a robot. A similar system was built by Hagita et al. [37], which supports wearable devices and is used mainly for sensing. The information infrastructure of these two systems is composed of reconfigurable modules that make the architectures scalable. However, they do not have middleware that supports consistent and simple application programming and management. The work closest to ours is the system built by Lee et al. [38], which used Jini as an architecture for supporting robots in a home network environment. However, their device is a mobile robot and involves no improvement to the original Jini model.

In terms of Jini systems, different kinds of extension can be found in the literature that aim at improving fundamental functionalities, including the support for group communication [39], self-adaptive leasing [40], and authentication of ad hoc services [41]. Cheng [42] has developed a service advertisement and discovery protocol for ad hoc communications and collaboration in ubiquitous computing environments. It does not, however, take into account the overall condition of the network and the interaction between the environment and the user in an intelligent manner in the way our system does.

Some characteristics of Jini determine the advantages and disadvantages of this experimental platform. Jini is based on Java language only; therefore, it is independent on operating systems and hardware, as long as a Java Virtual Machine (JVM) is installed. The lookup service in the Jini system provides a convenient mechanism for service discovery and advertisement; however, a catastrophic result will occur if the lookup service fails to perform properly. Jini supports code mobility, so security is an important issue. Currently, our system is implemented in a local area network and system security is limited to SSL signal encoding and user identification using database information. Some research on Jini security [41] provides useful insights into the future development of our architecture, such as the systems developed by Krone et al. [43] and Eronen and Nikander [44]. The former is an authentication and authorization architecture that uses secure communication channels and digital signatures. The latter is an architecture that separates the Java access permissions of Jini clients, service proxies, and services, while allowing for natural delegation of Java permissions between Jini-enabled devices.

2.7 CONCLUSION

In this chapter, we have introduced the concept of the wearable robot. We consider wearable robots to be distributed portable devices that can support intelligent interactions between the users and other networked entities. We have also developed a service-based architecture using Jini to enable the flexible and reconfigurable connection between the interacting components in the distributed network. We have extended the original Jini network model in three different ways that allow the system to prioritize tasks using priority queues in the task coordinator service, automatically download proxies to the relevant client based on the conditions of related parties, and automatically deliver service based on contextual information.

REFERENCES

1. A. Agah and K. Tanie, "Taxonomy of research on human interaction with intelligent systems," in *Proc. IEEE Int. Conf. on Systems, Man and Cybernetics*, Vol. 6, pp. 965–970, 1999.
2. K. Hinckley, et al., "Sensing techniques for mobile interaction," in *Proc. of ACM User Interface Software and Technology, CHI Letters*, Vol. 2, pp. 91–100, 2000.
3. A. Dey, "Providing architectural support for building context-aware applications," PhD Thesis, Georgia Institute of Technology, Atlanta, 2000.
4. T. E. Starner, "Wearable computing and contextual awareness," Ph.D. Thesis, Massachusetts Institute of Technology, Cambridge, MA, 1999.

5. G. S. Welling, "Designing adaptive environment-aware applications for mobile computing," Ph.D. Thesis, The State University of New Jersey, 1999.

6. S. Mann, "Telepointer: Hands-free completely self-contained wearable visual augmented reality without headwear and without any infrastructural reliance," in *Proc. 4th Int. Symp. on Wearable Computers*, pp. 177–178, 2000.

7. R. W. Picard, *Affective Computing*, MIT Press, Cambridge, 1997.

8. K. K. Lee and Y. Xu, "Real-time estimation of facial expression intensity," in *Proc. IEEE Int. Conf. on Robotics and Automation (ICRA)*, Taipei, Taiwan, Las Vegas, NV, 2003.

9. O. Kubitz, M. O. Berger, and R. Stenzel, "Client-server-based mobile robot control," *IEEE/ASME Transactions on Mechatronics*, Vol. 3, No. 2, pp. 82–90, June 1998.

10. R. C. Luo, T. M. Chen, and C. Y. Hu, "Networked intelligent autonomous mobile robot: issues and opportunities," in *Proc. IEEE Int. Symp. on Industrial Electronics*, Vol. 1, pp. 7–13, 1999.

11. S. Fleury, M. Herrb and R. Chatila, "GenoM: a tool for the specification and the implementation of operating modules in a distributed robot architecture," in *Proc. 1997 IEEE/RSJ Int. Conf. on Intelligent Robots and Systems*, Vol. 2, pp. 842–849, 1997.

12. S. Park, J. Choi, S. Baeg, M. Jang, G. Lee, and Y. Lim, "Message-based agent communications in a tightly coupled multiagent system," in *Proc. 4th Golden West Int. Conf. on Intelligent Systems*, pp. 194–198, San Francisco, CA, 1995.

13. D. W. Forslund, et al., "The importance of distributed, component-based healthcare information systems: the role of a service-based architecture," in *Proc. 14th IEEE Symp. on Computer-Based Medical Systems*, pp. 79–82, 2001.

14. R. Agrawal, R. Bayardo, D. Gruhl, and S. Paparimitriou, "Vinci: A service-oriented architecture for rapid development of web applications," in *Proc. 10th Int. World-Wide Web Conf.*, May 2001.

15. A. Diagne and M. P. Gervais, "Building telecommunications services as qualitative multi-agent systems: the ODAC project," in *Proc. IEEE Globecom'98*, Sydney, Australia, Nov. 1998.

16. R. D. Schimakat, W. Blochinger, and C. Sinz, "A service-based agent framework for distributed symbolic computation," in *Proc. 8th Int. Conf. on High Performance Computing and Networking, (HPCN Europe 2000)*, Amsterdam, The Netherlands, pp. 644–656, 2000.

17. A. Trigila, K. Raaiikainen, B. Wind, and P. Reynolds, "Mobility in long-term service architectures and distributed platforms," *IEEE Personal Communications*, Vol. 5, No. 4, pp. 44–55, Aug. 1998.

18. W. K. Edwards, *Core Jini*, The Sun Microsystems Press, 2001.

19. L. A. Zadeh, "Fuzzy logic = Computing with words," *IEEE Transactions on Fuzzy Logic*, Vol. 4, No. 2, pp. 103–111, May 1996.

20. S. Ghosh and Q. Razouqi, "A survey of recent advances in fuzzy logic in telecommunication networks and new challenges," *IEEE Transactions on Fuzzy Systems*, Vol. 6, No. 3, pp. 433–447, Aug. 1998.

21. P. Bosc and E. Damiani, "Fuzzy service selection in a distributed object-oriented environment," *IEEE Transactions on Fuzzy Systems*, Vol. 9, No. 5, pp. 682–698, Oct. 2001.

22. R. Zhang and Y. A. Phillis, "Fuzzy control of queuing systems with heterogeneous servers," *IEEE Transactions on Fuzzy Systems*, Vol. 7, No. 1, pp. 17–26, Feb. 1999.

23. L. S. Cheung and Y. K. Kwok, "The design and performance of an intelligent Jini load balancing service," in *Proc. IEEE Int. Conf. on Parallel Processing*, pp. 361–366, 2001.

24. W. Pedrycz, "Why triangular membership functions?" *Fuzzy Sets Systems*, Vol. 64, pp. 21–30, 1994.

25. K. K. Lee, M. Yu, and Y. Xu, "Modeling of human walking trajectories for surveillance," in *Proc. IEEE/RSJ Int. Conf. on Intelligent Robots and Systems (IROS)*, 2003.

26. P. Huang, "Promoting wearable computing: A survey and future agenda," *Technical Report TIK-Nr.95*, Computer Engineering and Networks Laboratory, Swiss Federal Institute of Technology, Sept. 2000.

27. Y. Tsumaki, Y. Fujita, A. Kasai, C. Sato, D. Nenchev, and M. Uchiyama, "Telecommunicator: a novel robot system for human communications," in *Proc. 11th IEEE Int. Workshop on Robot and Human Interactive Communication*, pp. 35–40, 2002.

28. A. Schmidt, H.-W. Gellersen, and M. Beigl, "A wearable context-awareness component – finally a good reason to wear a tie," in *Proc. 3rd Int. Symp. on Wearable Computers*, San Fransico, CA, pp. 176–177, 1999.

29. P. Ljungstrand, S. Bjrk, and J. Falk, "The WearBoy: A platform for low-cost public wearable devices," in *Proc. IEEE Int. Symp. on Wearable Computers*, 1999.

30. T. Kato, T. Kurata, and K. Sakaue, "VizWear-Active: Distributed Monte Carlo face tracking for wearable active camera," in *Proc. Int. Conf. on Pattern Recognition (ICPR2002)*, Quebec City, Canada, Vol. 1, pp. 395–400, 2002.

31. W. Mayol, B. Tordoff, and D. Murray, "Designing a miniature wearable visual robot," in *Proc. IEEE Int. Conf. on Robotics and Automation*, Vol. 4, pp. 3725–3730, May 2002.

32. T. Starner, J. Auxier, D. Ashbrook, and M. Gandy, "The gesture pendant: A self-illuminating, wearable, infrared computer vision system for home automation control and medical monitoring," in *Proc. 4th Int. Symp. on Wearable Computers*, Atlanta, CA, IEEE Computer Society Press, pp. 87–94, 2000.

33. G. D. Abowd, "Software engineering issues for ubiquitous computing," in *Proc. 1999 Int. Conf. on Software Engineering*, pp. 75–84, May 16–22, 1999.

34. T. Kindberg and A. Fox, "System software for ubiquitous computing," in *Pervasive Computing*, Vol. 1, No. 1, pp. 70–81, Jan.–Mar. 2002.

35. B. Rhodes, N. Minar, and J. Weaver, "Wearable computing meets ubiquitous computing: reaping the best of both worlds," in *Proc. The 3rd Int. Symp. on Wearable Computers*, pp. 141–149, 1999.

36. J.-H. Lee, G. Appenzeller, and H. Hashimoto, "An agent for intelligent spaces: functions and roles of mobile robots in sensored, networked and thinking spaces," in *Proc. IEEE Conf. on Intelligent Transportation System*, pp. 983–988, Nov. 9–12, 1997.

37. N. Hagita, K. Kogure, K. Mase, and Y. Sumi, "Collaborative capturing of experiences with ubiquitous sensors and communication robots," in *Proc. IEEE Int. Conf. on Robotics and Automation*, Vol. 3, pp. 4166–4171, Sept. 14–19, 2003.

38. B.-J. Lee, H.-G. Lee, J.-H. Lee, and G.-T. Park, "New architecture for mobile robots in home network environment using Jini," in *Proc. IEEE Int. Conf. on Robotics and Automation*, Vol. 1, pp. 471–476, 2001.

39. A. Montresor, R. Davoli, and O. Babaoglu, "Enhancing Jini with group communication," in *Proc. Int. Conf. on Distributed Computing Systems*, pp. 69–74, 2001.

40. K. Bowers, K. Mills, and S. Rose, "Self adaptive leasing for Jini," in *Proc. 1st IEEE Int. Conf. on Pervasive Computing and Communications*, 2003.

41. C. Gehrmann and P. Nikander, "Securing ad hoc services, a Jini view," in *Proc. 1st Annual Workshop on Mobile and Ad Hoc Networking and Computing*, pp. 135–136, Aug. 11, 2000.

42. L. Cheng, "Service advertisement and discovery in mobile ad hoc networks," in *Proc. Workshop on Ad hoc Communications and Collaboration in Ubiquitous Computing Environments,* in conjunction with the ACM 2002 Conference on Computer Supported Cooperative Work, New Orleans, LA, Nov. 16–20, 2002.

43. O. Krone, T. Schoch, and H. Federrath, "Making Jini secure," in *Proc. 4th Int. Conf. on Electronic Commerce Research (ICECR-4),* Dallas, TX, Nov. 8–11, 2001.

44. P. Eronen and P. Nikander, "Decentralized Jini security," in *Proc. The Network and Distributed System Security Symposium,* San Diego, CA, Feb. 2001.

■■■■■■■ **CHAPTER 3**

Wearable Interface for Automatic Language Translation[1]

3.1 INTRODUCTION

In this chapter, we introduce intelligent glasses, which can automatically translate multiple languages in real time. We propose the concept of the system and demonstrate the advantages of the device over other existing systems. First, the system architecture and the functions of the components in the system are presented. We then focus on the most crucial technical component, text detection. We propose a novel algorithm based on the fundamental characteristics of all the characters in common use called a character intrinsic characteristic (CIC)-based text detection algorithm. We then show the effectiveness of the proposed methods and define some future studies.

Nowadays, with the development of mobile computing technologies, a new kind of human interface has been expected to greatly impact our daily life and social activities. A wearable interface, just as its name implies, can be worn directly. It could be like a real cap, a pair of glasses, a pair of shoes, a shirt, and so on. Many wearable computers are facilitated with telepresence, which is a sophisticated form of interactive remote control, and some sensors such as a camera, gyro, infrared emitter, ultrasonic emitter, and so on. Many kinds of wearable computers are developed all over the world. Some wearable computers have been successfully applied in military missions, scientific exploration, and daily life.

This chapter presents a novel wearable interface project called Intelligent Glasses. With this interface, real-time multilinguistic translation can be realized and will be useful to a tourist who does not know a local language. Let us imagine that a person is using the wearable translation interface during a foreign trip. The system can tell the wearer which building is a hotel, a restaurant, a bank, or a supermarket, which bus line should be selected, how to go to a destination according to the

[1]Reprinted, by permission, from Xi Shi, Yangsheng Xu, and Jianming Hu, "A Wearable Translation Robot," in *Proc. of IEEE International Conference on Robotics and Automation*, pp. 4411–4416, Barcelona, Spain, April 18–22, 2005. Copyright © 2005 by IEEE.

translations of road signs, or which course is suitable for his taste based on the translation of a menu. All these options can make the tourist's journey much more delightful. Moreover, the Intelligent Glasses can also be regarded as a real-time multilingual translation dictionary. When you wear it, you can read any books written in any language.

In this project, the most important and difficult problem is to detect and extract the texts. In this research area, a great deal of effort has been placed in text detection in videos and images. Hua et al. [1] proposed an objective and comprehensive performance evaluation protocol for video text detection algorithms, which included a positive set and a negative set of indices at the textbox level to evaluate detection quality. It may be effective and applicable for improving, selecting, and designing new text detection methods. Li et al. [2] presented a robust text tracking scheme from video, which is composed of two modules: a sum of squared difference (SSD)-based module to find the initial position, and a contour-based module to refine the position. Li and Li [3] introduced a convolutional neural network (CNN) to study the possibility of building pattern classifiers for text/picture segmentation and text detection problems. Shin et al. [4] made use of a support vector machine (SVM) for the texture classifier and illustrated that SVM has some advantages over the traditional neural network-based text detection method. LeBourgeois [5] presented a general robust optical character recognition (OCR) system designed for practical use and suitable for grabbing unconstrained gray-level images from a charge-coupled device (CCD) camera. They extracted gray-level features, which made the algorithm more reliable, especially under poor printing conditions or bad contrast digitization.

Furthermore, in this research area, the Sign Translation Interactive System Lab of Carnegie Mellon University has also done a lot of outstanding work. References 6–8 proposed some automatic text extraction algorithms based on edge detection and layout analysis. References 9 and 10 introduced some translation systems on the platform of the personal digital assistant (PDA). Moreover, there are many successful visual tracking and recognition systems, which have similar functions to our project. Watanabe et al. [11] introduced a translation system from Japanese to English. Keaton et al. [12] introduced a multimodal wearable computer system "SNAP & TELL," which can quickly identify objects encircled by a pointing gesture and give an audio narration.

In this chapter, we propose a new text detection method called the CIC-based text detection algorithm. It performs well even in an abominable environment. For example, it can extract a correct text even though a digital camera produces a very blurred image. At the same time, the speed of the system is so fast that the text information can be detected and extracted in real time.

The remainder of the chapter is organized as follows. In Section 3.2, the system architecture will be introduced and the function of each module will be described in detail. The CIC-based text detection algorithm will be introduced in Section 3.3 Moreover, image cutting, rotation, binarization, and real-time translation methods will be introduced in Sections 3.4 and 3.5, respectively. Finally, we summarize the chapter and propose some future work in Section 3.6

3.2 SYSTEM ARCHITECTURE

The system is made up of a small head-mounted camera, wearable computer, and head-mounted display, as shown in Figure 3.1. These three components collect, translate, and present information about the detected text. By introducing different OCR systems and translation systems, we can realize translation between different languages by means of a flexible configuration. In summary, the main purpose of this system is to create a wearable device using a digital camera and wearable computer technologies for the enhancement of vision and language understanding.

Since the traditional keyboard and mouse are not suitable for this kind of circumstance, communication with the user is one of the most important problems for a wearable interface design. In this project, a wearable camera is used as an effective input sensor on the headset through which the system can collect information from the environment. A head-mounted display is the information display device in this system, which provides a visual interface for the wearer. The wearer can see the translated words or phrases via this device.

The wearable computer is the key component in this system. The images obtained by the camera are transmitted to this part continuously, and the text information in these images are detected and extracted by a text detection algorithm. The extracted text is then input into the cutting and rotation module, which can normalize the gradient text to a uniform shape. Furthermore, the binarization module can get rid of the background and preserve the text only. The binarized image is then input into the OCR system, which can transform the image into recognized characters or text. A corresponding translation system can translate the characters or text into the wearer's native language. Finally, the translated text is output and displayed on the head-mounted display.

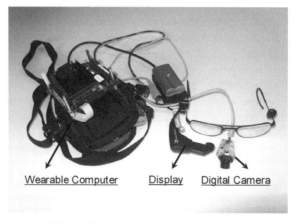

Figure 3.1. The three main components.

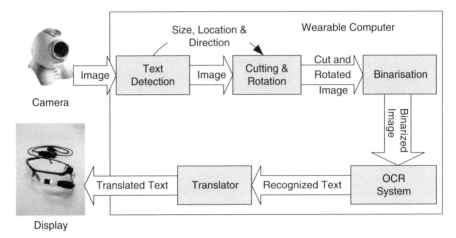

Figure 3.2. Detailed system architecture.

The detailed system architecture of the wearable interface is shown in Figure 3.2. A PC104 system produced by the Digitallogic Company is applied in this project. The main configuration of this wearable computer is as follows:

- CPU: Pentium III 800
- Memory: 256M
- Hard-disk: 1 G

The OCR system is an open source system, called GOCR for Linux, which can be downloaded from the following website: http://jocr.sourceforge.net/.

MicroOptical SV3 is introduced to this system as a head-mounted display. The specifications of this product are as follows:

- Display format: 640 × 480, 60 Hz refresh rate, landscape
- Color depth: 6 bit (64 colors)
- Input signal: VGA 60 Hz
- Focus range: Adjustable focus from 2 to 15 feet

3.3 TEXT DETECTION ALGORITHM

3.3.1 Demands of Text Detection Algorithm

As we know, the text in an image can be effectively used only after successful recognition. Before recognition, it is necessary to preprocess the image. Text detection is one of the most important approaches. The quality and speed of text detection both have a great impact on the performance of the whole system. To extract the text from the image obtained by the wearable camera, the algorithm in this project must satisfy the following requirements:

1. The speed of the algorithm must be fast to ensure that the system can run in real time; i.e., once the camera has caught the image, the translated words should be revealed in the display in a short time.
2. The algorithm should adapt complicated backgrounds and luminosity.
3. The algorithm should be suitable to any orientation of characters, including horizontal, vertical, gradient, or even in a curved surface.
4. Since the wearable camera is always in motion, the text detection and extraction algorithm should be adaptive for the vague image captured by a camera in motion or out of focus.
5. As the location information of a character will be used in other modules, the algorithm should minimize the noise.

3.3.2 Intrinsic Characteristic of a Character

In this section, we propose a new text detection algorithm called a CIC-based text detection algorithm. The idea of this method is based on the following assumptions:

Assumption No. 1: Each character in almost all languages is composed of short line segments, as shown in Figure 3.3.

Assumption No. 2: Within one character, the widths of line segments are similar.

Assumption No. 3: Suppose a character can be recognized in an image; the gray level of the character is different from that of the background.

Assumption No. 4: A line segment in a character should not be too long compared with its width.

Assumption No. 5: The gray level of a character is approximately even.

3.3.3 CIC-Based Text Detection Algorithm

The CIC-based text detection algorithm is introduced as follows.

1. *Gray-Level Difference Curve*

   Suppose that the image obtained from the camera is composed of $M \times N$ pixels; the gray level of each pixel is denoted as $g(i, j)$, where $i = 0, 1, \ldots,$

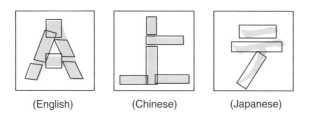

(English)   (Chinese)   (Japanese)

Figure 3.3. Characters are composed in line segments.

$M - 1$ and $j = 0, 1, \ldots, N - 1$. Also, $g(i, j) \in [0, 255]$. Then, an image can be stored in a gray-level matrix denoted as $\mathbf{G}$. Based on Assumption No. 3, we can obtain a gray-level difference matrix $\mathbf{G_\Delta}$, by

$$g_\Delta(i, j) = g(i + 1, j) - g(i, j) \tag{3.1}$$

where $i = 0, 1, \ldots, M - 2$ and $j = 0, 1, \ldots, N - 1$. Thus, for the jth row of the image, we can get the gray-level difference curve as shown in Figure 3.4. From the gray-level difference curve, we can see that there are a lot of pulses. Each of them represents a change of gray level. If the value of gray-level changes from small to large, then there is a positive pulse, and vice versa. When a line segment of a character goes along this scanning line, two pulses exist, one of which is positive and the other is negative.

2. *Finding Out the Pulse and Filtering the Noise*

There are three parameters for a pulse: start point, end point, and amplitude, which are denoted as p_s, p_e, and p_a, respectively. The width of a pulse is represented as p_w, which can be calculated by $p_e - p_s$. The sum of two adjacent pulses is the width of the line segment of one character. The algorithm for noise filtering is as follows:

- If there is only one positive or one negative pulse, then the pulse should be filtered.
- If the amplitude of a pulse is smaller than the threshold δ_a, it implies that the change of gray level is also small. According to Assumption No. 3, it does not belong to a character.
- If the width of the pulse is too large and the amplitude is low, then it shows that the gray level changes gradually, as shown in the second part of

Figure 3.4. Gray-level difference curve of one row in an image.

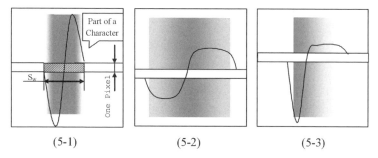

(5-1)   (5-2)   (5-3)

Figure 3.5. A sketch map of the noise filtering method.

Figure 3.5. According to Assumption No. 5, it is also not a part of a character.

- If the ratio of the amplitudes between two adjacent pulses is larger than threshold δ_w, it means that the gray level changes sharply. Also, according to Assumption No. 5, it does not belong to a character, as shown in the third part in Figure 3.5.

3. *Thread Clustering*

After noise filtering and scanning all the lines of the image, we can obtain a lot of "threads" as shown in Figure 3.6. The width of the thread is one pixel, and the length of the thread is the width of line of the character under investigation. Then, we should combine the threads into a line segment. The thread clustering includes six steps:

Step 1: *Initialization.* Let N_t denote the total number of threads, T_k ($k = 1, 2, \ldots N$) represent thread k, C_l ($l = 1, 2, \ldots, N_c$) denote the lth class, N_c denote the total number of classes (N_c may be different for different images), and $N_{t,l}$ denote the

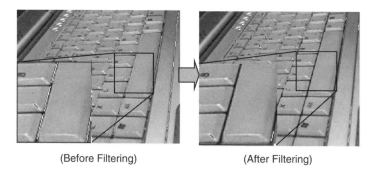

(Before Filtering)   (After Filtering)

Figure 3.6. Filtering the long line.

number of threads in Class C_l. Let $k = 1$ and $l = 1$. Then T_1 is a unique component of Class C_1.

Step 2: *Thread Clustering*: We denote the last thread in each class as F_{C_l}. Let $k = k + 1$. If $k > N_t$, then go to Step 3; otherwise judge the following condition:

$$|p_{s,k} - p_{s, F_{C_l}}| + |p_{e,k} - p_{e, F_{C_l}}| < \delta_t \tag{3.2}$$

where δ_t is a predetermined threshold value, which should be calibrated according to the experimental data. If condition (1) is satisfied, then T_k is classified into C_l and Step 2 is repeated; otherwise let $l = l + 1$ and T_k is classified as a new class.

Step 3: Let $l = 0$.

Step 4: Judge the number of threads in each class. Let $l = l + 1$; if $N_{t,l} < \delta_{nl}$, then eliminate this class and repeat Step 4, where δ_{nl} is the least number of threads in one class, which should be preset to 0; otherwise, go to Step 5.

Step 5: Judge the shape of each class. Calculate the average width (denoted as $\overline{w}_l$) and total length (denoted as TL_l) of class C_l according to Equations 3.3 and 3.4. Then calculate the ratio of TL_l and $\overline{w}_l$ (denoted as R_l). If $R_l > L$ (L is a preset value), then eliminate this class (according to Assumption No. 4); otherwise go to Step 6.

$$\overline{w}_l = \frac{1}{N_{t,l}} \sum_{i=1}^{N_{t,l}} (|p_{s,i} - p_{e,i}|) \tag{3.3}$$

$$TL_l = N_{t,l} \cdot I_p \tag{3.4}$$

where I_p is the width of one pixel.

Step 6: If $l > N_c$, then END; otherwise go to Step 4.

Using Step 5 illustrated above, the very long line can be filtered, which accelerates the system's running speed enormously as shown in Figure 3.6.

3.3.4 Combine Line Segments into a Character

The segment clustering algorithm is applied to combine the line segments into a character.

First, all line segments are labeled, and each segment is clustered into one class. Then, we find and record the center coordinate of each line segment. Moreover, we calculate the distance between any two segments. If a distance is less than a predefined value, then classify them into a new class. Finally, if two classes include the same segment, then cluster them into a larger new class.

Based on the above algorithm, the line segments can be integrated into characters.

3.4   IMAGE CUTTING, ROTATION, AND BINARIZATION

3.4.1   Image Cutting and Rotation

After detecting the characters, we should integrate some single characters into lines of text, which should be easier for the wearer to understand. We realize this function based on clustering. According to the differences of character size for different languages, we can predefine some thresholds. If the distance between some characters is less than the threshold, then they are clustered into one class. (In English, it may be a word or some words.) We use rectangles to confine the extracted words, as shown in Figure 3.7. To speed up the recognition process, only the part confined in rectangles is input into the OCR system.

As illustrated in Figure 3.7, there are some inclined rectangles. Before inputting them to the OCR system, they must be rotated (or normalized) to make sure that all the words are upright.

3.4.2   Image Binarization

After an image is processed by using the methods mentioned above, there are still some noises in the rotated rectangular subimage. There may be some parts of a character that are not detected. Therefore, there are two main tasks in this step. One is to build a character as much as possible. The other is to binarize the image. Image binarization involves the removal of all the background and noises and to keep the texts only in the final image to be input into OCR system. All the gray-level values of the background pixels are set to 255, i.e., they are white. On the contrary, all the

Figure 3.7. Confining the texts.

118	120	125	123	121	123	124	129	130	135	137	132
117	121	90	10	10	9	11	12	12	100	136	121
114	120	101	11	12	10	11	10	11	105	127	128
118	120	122	118	52	18	29	45	120	118	120	125
113	121	121	119	86	12	13	69	129	117	121	90
118	120	120	112	69	9	12	54	137	114	120	101
119	120	126	120	75	11	14	69	136	119	120	126
112	121	119	116	59	11	12	66	127	112	121	119
120	120	125	137	100	15	8	75	120	120	120	125
116	118	121	136	95	12	10	54	121	137	132	129
109	121	98	14	12	10	13	10	12	108	130	128
119	120	108	11	12	10	11	10	11	100	129	129
130	129	111	124	121	129	111	125	130	132	127	131

Figure 3.8. A sketch map of gray level for a detected character confining the text.

grey level values of the foreground pixels, which form the extracted texts, are set to 0, i.e., they are black.

In order to illustrate clearly, we present a sketch map (Figure 3.8), in which the value in each cell represents its average gray level.

For some selected zone, like the rectangle shown in Figure 3.8, two zones are marked with 18 and 29 not being found for some reasons. We first select the nearest adjacent threads and compute the average value of their gray levels, i.e., 41. Then, the average value can be regarded as a reference gray-level value for this rectangle zone. If the gray-level value of some certain zone is larger than the reference, then the gray level of this zone is set to 255. On the other hand, if it is less than the reference, the color of this zone is set into black; i.e., the gray level is 0. In this way, we can rebuild the character because the gray-level values of the two missed threads are all less than the reference value.

For the zone like rectangle No. 2, since there is no thread, that is, there is no character there, all the gray-level values in this zone are set to 255. We present an example as shown in Figure 3.9. It is a binarized subimage for the image displayed in Figure 3.7.

Figure 3.9. A subimage example after binarization.

Figure 3.10. An output interface shown in the headset display.

3.5 REAL-TIME TRANSLATION

The binarized subimages are first stored into a first-in–first-out buffer. Then, the OCR shell program reads them from the buffer and transmits them into the OCR system. Because of the processing speed restriction of the OCR system, the system will pause to deal with the images when the buffer is full.

The OCR system can extract the text information from the binarized image and change them into a string. The string is input to the translation system, which can be separated into words or phrases. Based on a language database, the translation system can translate the words and phrases into the other languages.

Finally, the translated words or phrases are shown in the head-mounted display (Figure 3.10).

3.6 CONCLUSION

In this chapter, we presented an Intelligent Glasses project that can detect, extract, and translate text into a user's familiar language in real time.

First, we described the system component and the main modules. Then, focusing on the most important module, text detection, we proposed a new algorithm called a CIC-based text detection algorithm. Moreover, image cutting, normalization, and binarization methods were also presented.

Currently, we have realized the mutual need for translation between Chinese and English. In the near future, we will design some other versions according to the different requirements of different users.

REFERENCES

1. X.-S. Hua, W. Liu, and H.-J. Zhang, "An automatic performance evaluation protocol for video text detection algorithms," *IEEE Transactions on Circuits and Systems for Video Technology*, Vol. 14, No. 4, pp. 498–507, Apr. 2004.

2. H. Li, D. Doermann, and O. Kia, "Automatic text detection and tracking in digital video," *IEEE Transactions on Image Processing*, Vol. 9, No. 1, pp. 147–156, Jan. 2000.

3. B. Li and B. Li, "Building pattern classifiers using convolutional neural networks," in *Proc. of International Joint Conference on Neural Networks (IJCNN '99)*, Vol. 5, pp. 1081–1085, 1999.

4. C. S. Shin, K. I. Kim, M. H. Park, and H. J. Kim, "Support vector machine-based text detection in digital video," in *Proc. of the 2000 IEEE Signal Processing Society Workshop*, Vol. 2, pp. 634–641, Dec. 11–13, 2000.

5. F. LeBourgeois, "Robust multifont OCR system from gray level images," in *Proc. of the Fourth International Conference on Document Analysis and Recognition*, Vol. 1, pp. 1–5, Aug. 18–20, 1997.

6. X. Chen, J. Yang, J. Zhang, and A. Waibel, "Automatic automatic detection of signs with affine transformation," in *Proc. of WACV2002*, Orlando, FL, Dec. 2002

7. J. Yang, X. Chen, J. Zhang, Y. Zhang, and A. Waibel, "Automatic detection and translation of text from natural scenes," in *Proc. of ICASSP2002*, Orlando, FL, May 2002.

8. J. Gao and J. Yang, "An adaptive algorithm for text detection from natural scenes," in *Proc. of IEEE Computer Society Conference on Computer Vision and Pattern Recognition*, Hawaii, Dec. 9–14, 2001.

9. J. Zhang, X. Chen, J. Yang, and A. Waibel, "A PDA-based sign translator," in *Proc. of ICMI2002*, Pittsburgh, PA, Oct. 2002.

10. Y. Zhang, B. Zhao, J. Yang, and A. Waibel, "Automatic sign translation," in *Proc. of ICSLP2002*, Denver, CO, Sept. 2002.

11. Y. Watanabe, Y. Okada, Y.-B. Kim, and T. Takeda, "Translation camera," in *Proc. of the Fourteenth International Conference on Pattern Recognition*, Vol. 1, pp. 613–617, Aug. 16–20, 1998.

12. T. Keaton, S. M. Dominguez, and A. H. Sayed, "SNAP & TELL: A multi-modal wearable computer interface for browsing the environment," in *Proc. of the 6th International Symposium on Wearable Computers (ISWC'02)*, pp. 75–82, Oct. 7–10, 2002.

CHAPTER 4

Intelligent Cap Interface for Wheelchair Control[1]

Many disabled people do not have the dexterity necessary to use a joystick or other hand interface for controlling a standard robotic wheelchair. In many cases, some of them suffer from diseases that can damage most of the nervous and muscular systems in the body but leave the brain and eye movement unimpaired. To this end, we developed an interface that enables a person to guide a robotic wheelchair by eye-gaze, using a minimal number of electrodes attached on the head of the user (Figure 4.1). The device can measure the electrooculographic potential (EOG) of the eye-gaze movement together with the electromyographic signals (EMG) from the jaw muscle motion. By coupling these simple actions, one can navigate a wheelchair solely by the eyes and the jaw, which provides an aid to mobility for people with a disability.

4.1 INTRODUCTION

Autonomous robotic wheelchairs represent an important class of autonomous mobile robots that is receiving increasing attention all over the world. This attention is due to the growing need for easier wheelchair mobility throughout a longer average life of people with a disability. Generally, the design of these powered wheelchairs follows the general principles, technologies, and methodologies of the other autonomous mobile robots. However, they must possess particular characteristics such as maneuverability, navigation, and safety that call for specialized designs.

Robotics wheelchairs require the integration of many areas of research, including vision, indoor/outdoor navigation, navigation with maps, reactive navigation, mode selection, sensor fusion, and user interfaces. However, most of these studies focus on the capability of autonomous navigation or, generally speaking, assistive navigation, by using vision or other sensing devices such as ultrasonic and infrared.

[1]Reprinted, by permission, from Cedric K. H. Law, Martin. Y. Y. Leung, Yangsheng Xu, and S. K. Tso, "A Cap as Interface for Wheelchair Control," in *Proc. of IEEE/RSJ International Conference on Intelligent Robots and Systems*, Volume 2, pp. 1439–1444, 30 Sept.–5 Oct. 2002. Copyright © 2002 by IEEE.

Amplifier

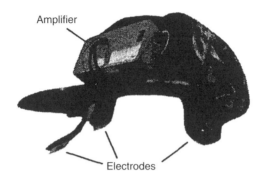

Electrodes

Figure 4.1. Interface on a cap.

The aim of intelligent wheelchairs is to provide wheelchairs with an ability to be controlled and navigated with minimal interactions with the users, in order to enhance the quality of service for people with a handicap. The most typical guidance device for a robotic wheelchair is a joystick. Other alternatives such as voice control, breath expulsion control, and guidance by head movements are also used.

Standard robotic wheelchairs have been in use for a long time; however, many people with a disability do not have the ability to control an ordinary powered wheelchair. An extreme disability such as severe cerebral palsy or amyotrophic lateral sclerosis (ALS) deprives them of the use of their limbs and certain muscles movement. Therefore, it is extremely difficult to express themselves through speech or bodily movement to command a robotic wheelchair. Some other advanced control alternatives for the robotic wheelchair users are worth exploring, which can meet the requirements for people with different levels of disability.

One of the solutions for the problem is to navigate by eye-gaze. In fact, there are several existing approaches to sense the eye movements. Most of these approaches involve the use of a camera to track some features of the eyes to determine where the user is looking, whereas another approach is to sense the EOG. In Refs. [1] and [2], a method to control and guide mobile robots by means of the ocular position (eye displacement into its orbit) is introduced. This method requires five electrodes to be placed around the eyes in order to detect the up and down, left and right movement of the eyes (Figure 4.2).

In our work, we have developed a device that measures the horizontal eye-gaze motions (looking to the left/right) and the jaw motion. By coupling these motions, the user can navigate a robotic wheelchair solely by the eyes and the jaw by using a minimal number of electrodes, which provides an aid to mobility for people with a disability.

This chapter is organized as follows: Section 4.2 discusses the bioelectrical signals in the human body. Section 4.3 describes the approach we are using in this work. Section 4.4 introduces the design of the interface. Section 4.5 is the experimental implementations, and Section 4.6 is the conclusion.

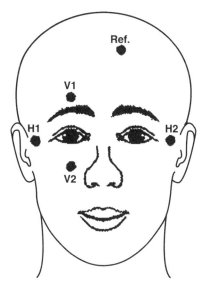

Figure 4.2. Traditional sensor placement method.

4.2 ELECTROMYOGRAPHY AND ELECTROOCULOGRAPHY

Electromyography (EMG) is a method of studying the activity of a muscle by recording action potentials from the contracting fibers. This method has been of value in the analysis of the actions of different muscles in the maintenance of posture and in the study of the physiology of motor unit activity. The signals that measured from the movement of the eyeball (ocular muscle) is typically known as EOG.

These bioelectric signals can be detected with surface electrodes that are easy to apply and can be worn for long periods of time while posing no health or safety risk to the user. Another advantage of this input method is that the hardware involved is comparatively simple, so that it is possible to design an EMG/EOG-based interface with minimal hardware in real-world applications.

4.3 APPROACH

The methods in Refs. [1–3] require five electrodes to issue the forward, backward, left, and right commands (Figure 4.2). In our approach, we demonstrated that three electrodes should be sufficient enough to guide a robotics wheelchair (Figure 4.3): one for the ground signal, and the remaining two for measuring 1) the eye-gaze at the horizontal direction and 2) the activation of the jaw muscles to trigger the forward/backward command. Table 4.1 compares the assignment of wheelchair commands in different approaches.

Figure 4.4 shows part of the muscles in a human's head. Region A is the location of ocular muscle, and EOG signals of the eyeball movement can be detected in this area.

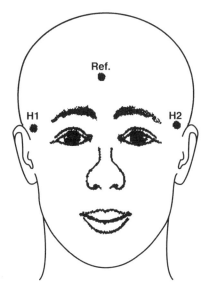

Figure 4.3. New sensor placement method.

Region B is called the temporal muscle, and this piece of fan-shaped muscle elevates the jaw and clenches the teeth. When the jaw moves, EMG signals will be detected at Region B. If we place a electrode at the overlapping area of Regions A and B, both signals produced by the eye movement and the jaw motion could be detected. The location is roughly the location of the temple between the eye and the ear on each side of the head (there maybe a slight variation from person to person).

Study in Ref. [4] showed that several problems would occur when this kind of eye movement interface is used. First of all, the eyes move jerky and rarely sit still, which results in inaccuracy in the performance of the device. Other factors like eyelid movement, variations in an individual's skin conductivity, and electrode placement also affect the results. To deal with the first problem, instead of using the interface as a direct substitute for a joystick, we treat the eye movements as "switches" to trigger different commands. For the latter problem, we reduced the number of electrodes to be used.

Table 4.1. Comparison of Using Different Approaches

Wheelchair Command	Electrodes Pair Involved	
	Trad. Method	Our Method
Turn Left	H1-H2	H1-H2
Turn Right		
Forward	V1-V2	H1-H2
Backward		

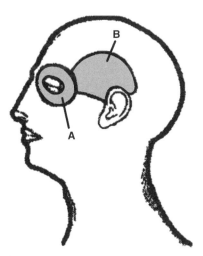

Figure 4.4. Muscles on the head.

We designed the input pattern when using this interface to command a robotic wheelchair (Table 4.2). Since people with a disability cannot turn back to look clearly at the scene behind the wheelchair, we suggested that it would be dangerous to allow the backward motion to be controlled by him/her. Therefore, we have disabled the backward command in our design. In fact, if the wheelchair system has been equipped with certain obstacles avoidance abilities, the user does not necessarily need to execute this command. As a summary, this approach has the following advantages:

- Stability and reliability increase
- Simple instrumentation
- Lower cost
- More comfortable to use

4.4 INTERFACE

4.4.1 Hardware

The electronic circuit is designed to detect the horizontal eye-gaze and the jaw motion. Three silver-silver chloride electrodes are applied on the surface of the

Table 4.2. The Input Pattern of Our Design

Command	Motion
Turn Left	Look to the left (L)
Turn Right	Look to the right (R)
Forward	Jaw motion (J)
Backward	-Disable-

skin to pick up the bioelectrical signal produced from the muscles. Since the signals are weak, we use an instrumentation amplifier as the pre-amplifier. This signal is a combination of a muscle movement (AC signal) and a potential difference of skin position (DC offset). We use two operational amplifiers to eliminate the DC offset and to amplify the AC signal further. The overall voltage gain is 72 dB.

To separate the signals generated by the eyes and the jaw, three more operational amplifiers are used. We can adjust the value of the potentiometers for the corresponding amplifier in order to match different muscle strength. Next, we use two monostable multivibrators to generate a standard transistor-transistor logic (TTL) signal format. We also use a logic circuit to prevent the resultant signals from appearing simultaneously. Two 3 V, 500 mAh Lithium coin cells are applied to drive the circuit; the life of the circuit is about 50 hours.

We also designed another circuit with LEDs to display the operation of the interface. One of the three big LEDs will turn on while the corresponding muscle movement is detected, and there are ten smaller LEDs to indicate the eye-gaze.

Finally, we integrated the whole hardware into a cap (see Figure 4.5) so that the electrodes can be installed/removed handily. All cables are hidden inside the cap, which makes the appearance better and prevents it from blocking the vision of the user.

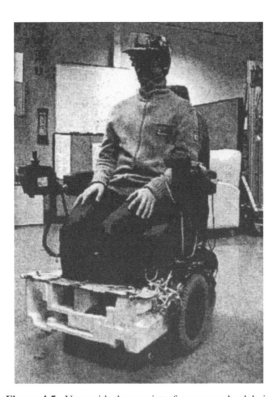

Figure 4.5. User with the cap interface on a wheelchair.

4.4.2 Implementation

Figures 4.6–4.8 show the bioelectrical signals we obtained from a user by using the three electrodes interface. The lower graph in each figure is the signal that directly picked up from the electrodes, whereas the upper one is the corresponding state of the wheelchair, which is driven by the user.

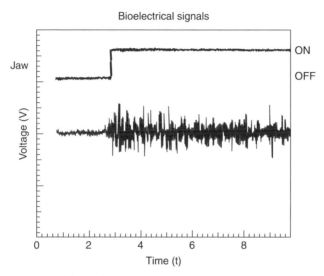

Figure 4.6. Activation of the JAW signal.

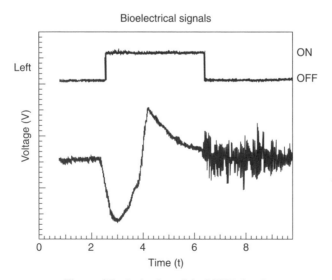

Figure 4.7. Activation of the LEFT signal.

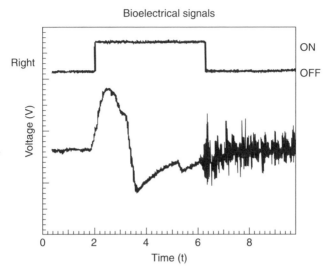

Figure 4.8. Activation of the RIGHT signal.

When the user is issuing a jaw motion command, both the electrodes at HI and H2 will detect the electrical activities from the jaw muscle, so that the resultant signal will oscillate at a very high frequency. When the user is looking to the left or right, a large potential difference will be detected at HI and H2: A negative change stands for the left, whereas a positive change stands for the right.

From the above results, we verify that the interface can detect three kinds of motions on the face: 1) looking to the left (L), 2) looking to the right (R), and 3) tightening the jaw (J). Since we have not applied any electrode along the vertical direction of the eye, eyelid movement such as blinking does not interfere with the signals in the H1–H2 electrodes pair. Next, we design the mapping between the states of the wheelchair and the actions produced by the users in Table 4.3. The difference between δ_{2-4} and δ_{5-6} is that the former one is to let the wheelchair turn slightly to the left/right ($\sim 20^{o}$) while it is moving ahead, which is suitable in fine tuning the path that is traveling; the latter one is to make the wheelchair rotate at its center, which is suitable in turning some large corners. When the wheelchair is moving, either it is turning or

Table 4.3. Control mapping

Wheelchair state	Symbol	L	R	J
Stop	δ_1	δ_5	δ_6	δ_2
Straight	δ_2	δ_3	δ_4	δ_1
Turn left	δ_3	δ_3	δ_3	δ_1
Turn right	δ_4	δ_4	δ_4	δ_1
Rotate left	δ_5	δ_5	δ_5	δ_1
Rotate right	δ_6	δ_6	δ_6	δ_1

moving straight ahead, the user can make a "STOP" whenever a jaw motion is generated. This can make sure that the user can stop the wheelchair motion actively to prevent any fault command from being made. The effects are demonstrated in Figure 4.7 and 4.8. In Figure 4.7, the user first looks to the left to trigger the "L" command; when he/she issues a jaw motion, the "L" state will be reset.

4.5 EXPERIMENTAL STUDY

Several experiments are conducted to show the maneuverability of the wheelchair when the interface is treated as the main input device. The wheelchair platform that we are currently using is TAO-6 from Applied AI Systems, Inc., Ottawa, Canada. Most of the electronics are housed in a protective box beneath the seat. The sensors, joystick, LCD display and keypad are connected directly to the electronics box. There are a total of 11 infrared sensors, 8 ultrasonic sensors, and 2 bumpers for local obstacles avoidance. The system uses the Motorola mc68332 32-bit micro-controller as the CPU. All sensor inputs and the user interface are handled by this unit and additional input/output boards.

The experiments are designed to meet the needs of the daily activities that a disabled person might perform in an indoor environment [Figure 4.9, such as passing through doorways (A–B), U-turning from a dead-end (B–C–B), and general navigation (C–D)]. Figures 4.10, 4.11 and 4.12 show the result of each experiment we have conducted. The sampling rate of the data is 64 Hz.

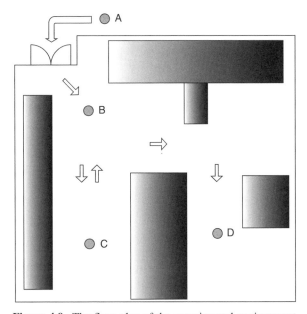

Figure 4.9. The floor plan of the experimental environment.

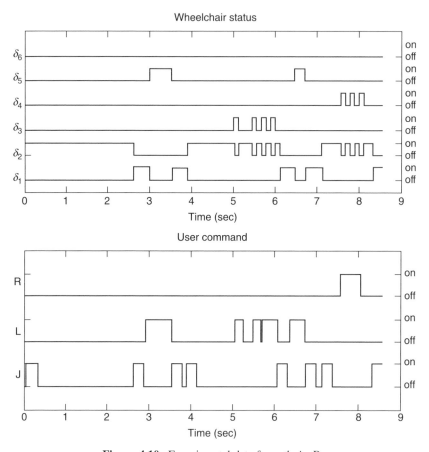

Figure 4.10. Experimental data for path A–B.

4.5.1 Doorways (A–B)

The task of passing through a doorway is an action that needs to be done everyday; whether the interface can enable the user to complete this task efficiently is a important concern. In Figure 4.10, the user is first issued a jaw command (J) to ask the wheelchair to move toward the doorway (δ_2). When the user has reached the doorway, he stops the wheelchair and issues a left command (L), which lets it rotate to the left (δ_5) until the heading is correct ($t \approx 4$). Then, the user is issued a "J" again in order to pass through the door. At $t = 6$, when the wheelchair has approached the block near the door, the user stops the wheelchair and rotates to the left to get away from that obstacle. At the end, the wheelchair stops at B. The whole experiment takes about 10 seconds to complete.

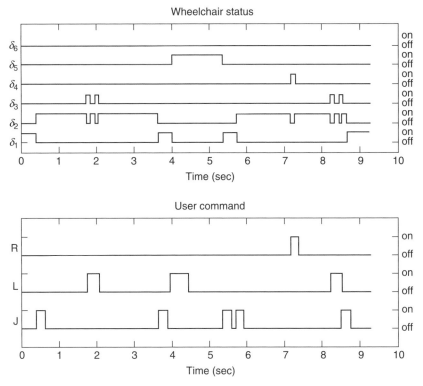

Figure 4.11. Experimental data for path B–C.

4.5.2   U-turning (B–C–B)

This task demonstrates how a user can turn back from a dead end. Figure 4.11 shows the result of this experiment. From $t = 0$ to $t = 4$, the wheelchair proceeds straight ahead to C. After it has reached C, the user commands it to rotate (δ_5) and then moves straight (δ_2) to head for B again. Due to the hardware limitation and disturbance on the floor, the wheelchair cannot move straight ideally, so the user needs to issue "L" or "R." to alter the path, for example, at $t \approx 2$ and $t \approx 7$.

4.5.3   General Path (C–D)

The last task is to travel from C to D; the result is given in Figure 4.12. It is interesting to note that the user does not need to stop the wheelchair to turn the first corner (at $t = 2.5$ to $t = 4$), only by executing a circular path for the wheelchair (δ_4). The above results show that the interface we developed is efficient enough to navigate a wheelchair. All the tasks were achieved successfully without getting hit on any wall or obstacle during the experiments. Results also show that the users can command the wheelchair to move or stop effectively. By providing enough training for the user to use the interface, the user can navigate the wheelchair freely as he/she wishes.

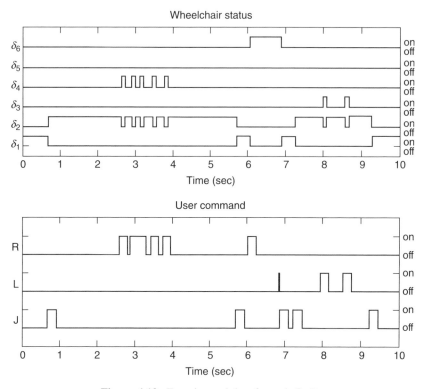

Figure 4.12. Experimental data for path C–D.

4.6 CONCLUSION

This research is aimed at developing a simple, low-cost, yet usable assistive robotic wheelchair system for disabled people, especially those with a limb disability. The interface that is presented in this chapter uses a minimal number of electrodes to navigate a robotic wheelchair by eye-gaze and simple jaw motion. With the reduction in the number of electrodes required, the hardware is simplified and the whole device can be fitted in an ordinary cap. The experimental results demonstrate that a person can guide the wheelchair within an indoor environment efficiently by using the interface we have developed. With more robust autonomous modules to be added on this system, we believe that the interface will be an excellent alternative input device other than the tradition joystick control for people with a disability.

This chapter presents a preliminary stage of this research. One disadvantage of our approach is that the performance is very sensitive to the electrode placement and skin conductivity of each individual. However, once the calibration is done at the beginning, the device will become very stable and reliable. As a next step, we are going to develop a wireless communication between the wheelchair and the device. We will also use the device to cope with the autonomous control module of the wheelchair to perform various navigation tasks.

REFERENCES

1. R. Barea, L. Boquete, M. Mazo, E. Lopez, and L. M. Bergasa, "EOG guidance of a wheel-chair using neural networks," in *Proc. of the IEEE 15th International Conference on Pattern Recognition*, Vol. 4, pp. 668–671, 2000.

2. H. Yanco and J. Gips, "Preliminary investigation of a semi-autonomous robotic wheelchair directed through electrodes," in *Proc. of the Rehabilitional Engineering Society of North America Annual Conference*, RESNA Press, pp. 414–416, 1997.

3. J. Gips and P. Olivieri, "EagleEyes: An eye control system for persons with disabilities," in *Proc. 11th International Conference on Technology and Persons with Disabilities*, Mar., 1996.

4. R. J. K. Jacob, "Eye movement-based human-computer interaction techniques: Toward non-command interfaces," *Technical Report*, Human-Computer Interaction Lab., Naval Research Laboratory, Washington, DC.

FURTHER READING

R. A. Cooper, "Intelligent control of power wheel-chairs," *IEEE Engineering in Medicine and Biology Magazine*, Vol. 14, No. 4, pp. 423–431, Jul.-Aug. 1995.

L. A. Fried, *Anatomy of the head, neck, face, and jaws*, Lea & Febiger, Philadelphia, PA, 2nd ed., 1980.

Y. Kuno, T. Yagi, and Y. Uchikawa, "Development of eye pointer with free head-motion," in *Proc. of the Annual Int. Conf. of the IEEE Engineering in Medicine and Biology Society*, Vol. 4, pp. 1750–1752, 1998.

S. H. Kwon and H. C. Kirn, "EOG-based glasses-type wire-less mouse for the disabled," in *Proc. of the First Joint BMES/EMBS Conference*, Vol. 1, pp. 592, 1999.

G. Norris and E. Wilson, "The Eye Mouse, an eye communication device," in *Proc. of the IEEE 1997 23rd North-east, Bioengineering Conference*, pp. 66–67, 1997.

R. C. Simpson and S. P. Levine, "Adaptive shared control of a smart wheelchair operated by voice control," in *Proc. of the 1997 IEBE/RSJ Int Conf. on Intelligent Robots and Systems, IROS '97*, Vol. 2, pp. 622–626, 1997.

H. Yanco, "Integrating robotic research: a survey of robotic wheelchair development," in *Proc. AAAI Spring Symposium on Integrating Robotic Research*, Stanford, CA, Mar. 1998.

H. Yanco, A. Hazel, A. Peacock, S. Smith, and H. Wintermute, "Initial report on Wheelesley: A robotic wheelchair system," in *Proc. Workshop on Developing AI Applications for the Disabled, Int. Joint Conf. on Artificial Intelligence*, 1995.

CHAPTER 5

Intelligent Shoes for Human–Computer Interface[1]

5.1 INTRODUCTION

Recently, much research in the integration of computing and sensor technologies into human clothing has been initiated. Nonetheless, one area of wearable devices has remained relatively unexplored. This is the design and implementation of sensor- and computer-equipped intelligent shoes. Some work for human gait has focused on foot parameter detection, such as temperature, humidity, heel-off time, gait velocity, and so on. There has been little work in analyzing the foot signal despite a lot of useful data that can be extracted from the shoe. The on-going miniaturization revolution in electronics, sensor, and battery technologies, driven largely by the cell phone and handheld device markets, has made possible an intelligent shoe implementation. Along with these hardware advances, progress in human data modeling and machine learning algorithms have also made possible the analysis and interpretation of complex, multichannel sensor data.

The data glove [1], a standard input device in virtual reality applications, is used for capturing finger and wrist motion. It interprets data from different sensors and lets the user be involved in a virtual environment. Hand motion is much more investigated than foot motion in that human hands are more flexible and can perform more complicated operations.

Inspired by the successful applications of the data glove, we now investigate the possible applications of a shoe-based sensor interface. Much useful data exists that

[1]Reprinted, by permission, from Weizhong Ye, Yangsheng Xu, and Ka Keung Lee "Shoe-Mouse: An Integrated Intelligent Shoe," in *Proc. of IEEE International Conference on Intelligent Robot and Systems*, pp. 1947–1951, Edmonton, Canada, Aug. 2–6, 2005. Copyright © 2005 by IEEE; from Meng Chen, Bufu Huang, Ka Keung Lee, and Yangsheng Xu, "An Intelligent Shoe-Integrated System for Plantar Pressure Measurement," in *Proc. of the IEEE International Conference on Robotics and Biomimetics*, pp. 416–421, Kunming, China, Dec. 17–20, 2006. Copyright © 2006 by IEEE; and from Bufu Huang, Meng Chen, Weizhong Ye, and Yangsheng Xu, "Intelligent Shoes for Human Identification", in *Proc. of the IEEE International Conference on Robotics and Biomimetics*, pp. 601–606, Kunming, China, Dec. 17–20, 2006. Copyright © 2006 by IEEE.

can be extracted from the shoe, especially information regarding motion when people perform different actions such as walking, standing, running, loading, and unloading. By analyzing the information given by the sensors inside the shoes, we can interpret the status of a person by his feet.

Some initial intelligent shoe systems have been prototyped. In particular, Morley et al. [12] have developed an electronic system for a shoe that monitors temperature, pressure, and humidity; however, only a hardware design is presented and there is no discussion on how the collected data are analyzed. Skelly and Chizeck [3] have presented a rule-based gait event detector with fuzzy logic and have concluded that two force sensitive resistors (FSRs) per insole are sufficient for gait event detection during walking. However, the robustness to nonwalking activities (shifting the weight from one leg to the other) is questionable. Williamson and Andrews [4] have reported excellent detection reliability by using three accelerometers attached to the shank and a machine-learning algorithm to detect the transitions among five gait phases during walking in real time, but no results have been presented for the use of this system with an functional electrical stimulation (FES) system. The Salisbury Group (U.K.) [5] has administered to several hundred patients the Odstock Dropped Foot Stimulator (ODFS). The foot switch indicates the heel-off and the heel-strike phases. The subjects learn to keep the foot switch pressed when they stop walking in order to avoid false stimulation triggers.

Moreover, various gait systems have been built with more functions and more signals for motion research. Paradiso et al. [6] have developed a wearable computer system for digital music that consists of a pair of instrumented sneakers for interactive dancing. In this work, the researchers created a few musical mappings with the shoes for computer-augmented dance. Also, Morris and Paradiso [7] have developed a compact, wireless, and wearable sensor package that is designed to provide continuous and real-time monitoring of gait for clinical applications. Pappas et al. [8] have proposed to analyze human gait patterns by using sensors attached to shoes; their system can distinguish walking from loading, unloading, or sliding of the foot. Finally, Kale et al. [9] have introduced a view-based system to recognize humans based on their gait. A hidden Markov model is used to capture the gait information in a high-dimensional image feature. Many researchers have analyzed foot motion through a set of heuristic rules. This approach, however, is only effective for simple motion patterns. Moreover, action recognition is typically based only around that of the simple motions of single individuals.

In this chapter, we develop the sensor-integrated shoe as an information acquisition platform to sense foot motion. The system is small, portable, and wearable. First, we integrated all the sensors and circuits fully inside the shoe without adding much weight into the original shoe. It is easy to use, and users will notice little difference between their normal shoes and the proposed shoes. Second, we built a hardware platform to collect data from the shoe. This platform is programmable, easily scalable, and easy to be integrated to the other applications. The platform is mainly composed of four parts, including a sensing module, a computing module, a wireless communications module, and a data visualization module. Based on this platform, we develop a novel input device called the Shoe-Mouse, which can be used by people who have difficulties in using their hands to operate computers or devices. We applied the intel-ligent shoes to three successful tasks, including the Shoe-Mouse, plantar pressure measurement, and human identification. Details of the application performance can be found in Section 5.3.

5.2 HARDWARE DESIGN

The proposed shoe-based information gathering platform consists of four subsystems. Figure 5.1 shows the architecture of our proposed platform.

Subsystem 1 is for sensing the parameters inside the shoe. A variety of sensors are installed inside the package, including force sensors, bend sensors, accelerometers, and so on. For ease of use, we limit the size of each device to as small as possible. Existing Micro ElectroMechanical Systems (MEMS) technology makes it possible to integrate all the sensors and circuits inside a small module.

Subsystem 2 is for gathering data from the sensors inside the shoe and sending the processed data to the wireless module. The processing power of the microprocessor is limited. It can only perform some simple calculations such as counting and averaging.

Subsystem 3 is for wireless communication. This communication system is com-posed of an emitter and a receiver. The receiver is for collecting the data from the circuit described in subsystem 2, whereas the emitter is for sending the data to the host computer for further analysis.

Subsystem 4 is for the visualization of the data. The received data are stored and displayed in real time on the screen of the host computer as a visual interface. This visual interface can be used for additional applications.

The prototype of the Intelligent Shoe is shown in Figure 5.2.

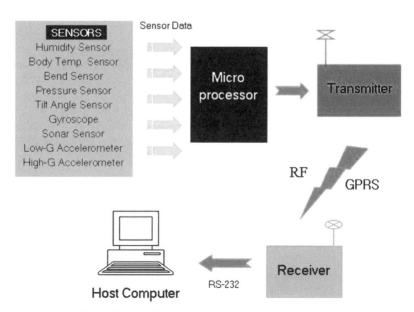

Figure 5.1. Outline of the hardware design and sensor.

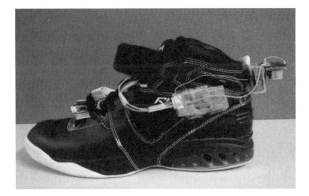

Figure 5.2. The prototype of the Intelligent Shoe.

5.2.1 Sensing the Parameters Inside the Shoe

To detect the important parameters and features of gait, a variety of sensors are installed in the shoes, including force sensors, bend sensors, switch sensors, accelerometers, gyroscope sensors, and ultrasonic sensors. Existing MEMS technology makes it possible to integrate all the sensors and circuits inside a small module.

Force-sensitive resistors (FSRs) and switch sensors are selected to detect the gait timing and pressure parameters. The force sensors operate with a voltage source and a fixed resistor to produce a voltage that changes with the applied forces. Although several FSRs cannot detect force distribution of a whole foot, we can get the main force feature and gait timing parameters for identification training.

One bend sensor is selected for gait flexion detection. The bend sensor is put in the insole under the big toe and heel. The resistance of the bend sensor changes as it is bent, which can provide information about flexion between the toe and the heel. The output of the bend sensor also contains rich information about human motion, especially loading and uploading of feet.

We select three single-axis gyroscope sensors and a three-axis accelerometer to detect motion orientation of the foot. Three single-axis MEMS gyroscope sensors (ENJ-03J Murata, Murata Manufacturing Co. Ltd., Kyoto, Japan) are mounted. As a miniature vibrating-read gyro, it uses piezoelectric material to sustain vibration, while taking advantage of Coriolis forces to measure angular rate. Each gyro sensor can test one angular rate in one direction; thus, we can measure yaw, roll, and pitch of shoe motion. Also, a three-axis MEMS accelerometer (MMA7260Q Freescale Semiconductor Inc., Austin, TX) is mounted, which can detect the acceleration motion of shoe in three dimensions. The gyroscope sensors and accelerometer can detect three-dimension rotation parameters and three-dimension acceleration parameters, which can be called the inertial measurement unit (IMU).

On the other hand, one ultrasonic sensor is added to measure the height between the shoe and the ground.

5.2.2   Gathering Information from the Sensors

This subsystem is mainly composed of a processor circuit board. The original analog signal generated by the sensors is transmitted directly to the ADC channels of the micro-processor (ATMega 8535, Atmel Corporation, San Jose, CA). After A/D transform, the digital signal is passed to a wireless communication module through the transmission data (TXD) port for transmission to a PC for data analysis and visualization. A micro battery cell is also added to serve as power supply. The circuit board is small, and it can be easily put into the heel of the shoe so that users will barely notice the difference between normal shoes and intelligent shoes.

5.2.3   Wireless Communication

This subsystem is for transferring the data from the shoe to the host computer. Many foot-based gait analysis systems do not use wireless systems as these will introduce many transmission errors that make the analysis result unstable. In our system, the size of the data are relatively small and it is possible to use a wireless system. We select the GW100b wireless communication module (Unitel Pty. Ltd., Taiwan), which has a 192,000 bps transmission speed and low power consumption (less than 10 mW). With an embedded microprocessor, the GW100b can realize forward error correction (FEC), which observably reduces transmission error and improves wireless communication reliability. At the same time, with the same power consumption and error rate requirement, GW100b can transmit data for further distances with the help of the FEC rather than with other wireless communication modules without the FEC function. The wireless communication process flow is shown in Figure 5.3.

5.2.4   Data Visualization

As this platform is designed for general applications, we display all parameters measured from the shoe. The host computer gets the data from the wireless receiver

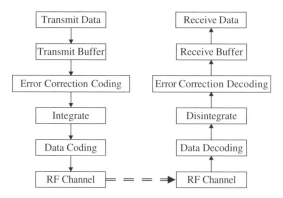

Figure 5.3. Wireless communication process flow.

Figure 5.4. Different gait events by animation.

via RS232. Different functions for visualizing the data from different sensors are developed and compacted. As an example, Figure 5.18 describes the real-time signals of IMU. We can also visualize a walking person by animation (in Figure 5.4), which is mapped from different gait events of the person. All of the above provide a friendly interface to display the data of the sensors obtained from the shoe.

5.3   THREE APPLICATIONS OF THE INTELLIGENT SHOES

In the following, section we will discuss three possible applications of the intelligent shoes, including 1) intelligent shoes for human–computer interface: shoe-mouse; 2) intelligent shoes for pressure measurement; and 3) intelligent shoes for human identification.

5.3.1   Intelligent Shoes for Human–Computer Interface: Shoe-Mouse

5.3.1.1   Motivation. The Shoe-Mouse, as we call it, aims to use the shoe as a mouse to operate computers or devices. The Shoe-mounted mouse, as far as we know, has never been investigated before. One important reason is that the human foot is not as flexible as the human hand. Our motivation for creating the Shoe-Mouse comes from the following inspirations.

Including the ordinary mouse and keyboard, many input devices have been invented to help people interact with computers more easily. Most of them are used by hands and arms, which may cause injuries related to shoulders and arms after prolonged use.

Persons who have difficulties in operating computers using their hands will benefit a lot from this invention, as they can use their feet to control computers.

As we can see from the principle of other devices such as the ordinary mouse and the data glove, information of motion can be applied to map the intention of humans to the computer. Thus, we think that by integrating our developed platform for modeling the motion of the foot, it is possible to gather information from the shoe as a mouse.

5.3.1.2   Analysis of Signals from Sensors Inside the Shoe. Six parameters are selected and measured: 1) the force sensor installed at the toe; 2) the

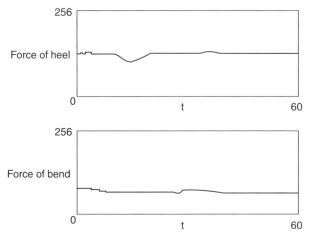

Figure 5.5. Visulization of data.

force sensor installed at the left side of the heel; 3) the force sensor installed at the right side of the heel; 4) the X-axis of the accelerometer at the heel; 5) the Y-axis of the accelerometer at the heel; and 6) the degree of bend from the bend sensor. Figure 5.5 shows the sample data taken from the Intelligent Shoe when a person wearing this shoe is walking at a speed of around 5 km/h.

It can be deducted from the figure that the data from the sensors can be easily mapped into the person's walking motion. It can be easily explained as follows: For time t from 0 s to 5 s, the force at the heel is about 50 units, which reflects the weight of the person, as a heavy person will have a larger weight on the heel of his shoe. For time t from 5 s to 7 s, the value of force at the heel becomes smaller than the original value, which means that a person is lifting his leg. For time t from 7 s to 9 s, the value of force at the heel returns to the original level again. At some key points, the value of the force is bigger than the original level, which shows that the person's foot is touching the ground. Other parameters, such as the force at the toe, and the values of the accelerometer sensors, can also be easily explained according to a person's motion.

5.3.1.3 Smoothing the Curve of the Data. In general, errors will be unavoidably introduced into the system. There are two kinds of errors. One is the error introduced from the wireless module, which can be improved by using more accurate modules. The other is introduced from the abnormal contact between the shoe and the human foot, which is the output of the sensors that do not reflect the person's intention and that should thus be deleted. These two errors will cause some abnormal peaks in the output waveform.

Here, exponential smoothing [7] is applied to minimize the effect of the abnormal peaks without affecting the performance.

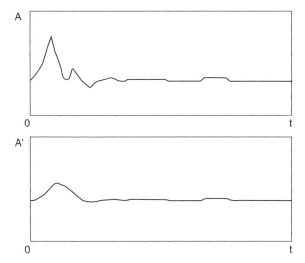

Figure 5.6. Result of exponential smoothing.

The principle of exponential smoothing can be described as follows:

$$S_t = \alpha * y_t + (1 - \alpha) * (S_{t-1} + b_{t-1}) \quad 0 \le \alpha \le 1 \tag{5.1}$$

$$b_t = \gamma * (S_t - S_{t-1}) + (1 - \gamma) * b_{t-1} \quad 0 \le \gamma \le 1 \tag{5.2}$$

In Equations 5.1 and 5.2, S, α, and γ are the parameters to be decided by the user. y is the original data, and b is the processed result. As can be observed in Figure 5.6, the processed wave is smoothed while preserving the original basic configuration.

5.3.1.4 *Mapping Motion of the Foot to the Motion of a Mouse-Cursor on the Screen.* After analyzing the sensor data from the shoe and applying exponential smoothing to remove some noise, the motion of the foot can be mapped into the motion of a mouse-cursor on the screen. The motion of the user's shoe can cause the motion of the mouse-cursor on the screen.

Inspired by the ordinary mouse (used by hand), we divide the motion of the mouse into four classes: 1) the mouse movement in two dimensions, that is, the $x-y$ position of the mouse-cursor on the screen; 2) the single click of the left button; 3) the single click of right button; and 4) the double click of the left button. Usually, the former three functions are enough to control a computer.

Based on the shoe-mounted data gathering platform, we use the following mapping methods to achieve the goals:

1. *The Use of the Accelerometer to Produce a Motion of the Mouse-Cursor.* The accelerometer used here (ADXL202E, Analog Devices, Norwood, MA) can output x-axis acceleration and y-axis acceleration. When the shoe is moved, the acceleration

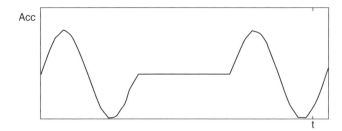

Figure 5.7. Output of x-axis acceleration when the shoe is following a rectangular route on a plane.

of the shoe is equivalent to that of the accelerometer. Figure 5.7 displays a sample output waveform of the *x*-axis acceleration when the shoe is moved back and forth according to a defined square route.

The output waveform of *x*-axis acceleration when the shoe is moving can be observed in Figure 5.8. It is difficult to use this signal to map the motion of the foot to the motion of the mouse-cursor. Fortunately, ADXL202E can sense the tilt angle in two directions. By using these data, we can easily map the motion of the foot into the motion of a mouse-icon on the screen. Figure 5.8 displays the output of the waveform of the *x*-axis and *y*-axis acceleration when the shoe is moved with different tilt angles.

2. *The Use of the Force Sensors at the Toe to Produce the Single Click of the Left Button.* The force sensor at the toe is naturally mapped into a single click. When a user has clicked the shoe using his toe, a force will be produced from the force sensor installed between the toe and the shoe. This force will persist for about 0.5 s. The average force during this time is computed and then compared with a threshold that can be calculated from some testing samples. If the force is bigger than the threshold, the function of a single click of the left button of the mouse takes effect.

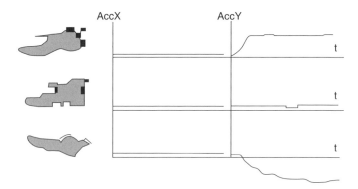

Figure 5.8. Motion of foot and its corresponding acc output.

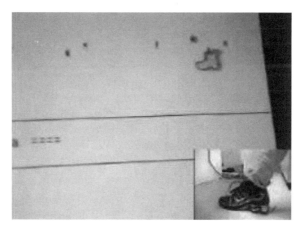

Figure 5.9. A picture captured from video demonstration.

3. *The Use of the Two Force Sensors at the Heel to Produce the Single Click of the Right Button.* The principle in this action is similar to that of action 2. The difference is that here we use two force sensors to judge whether the user has pressed the right button of the mouse. Simple fusion methods such as AND can be implemented in this stage. All three functions above make the control of the mouse easier. Figure 5.9 shows a picture that is captured from one of our video demonstrations.

5.3.1.5 Experiments and Performance.
Two experiments were conducted to evaluate the performance of the Shoe-Mouse: 1) measurement of the force sensor installed at the toe of the shoe; and 2) measurement of the foot movement on the $x-y$ plane to determine the trajectories of the movement as sensed by the motion sensor. The first experiment was performed to evaluate the efficiency of the left single click of the Shoe-Mouse, and the second experiment was performed to evaluate the efficiency of the movement of the Shoe-Mouse because the "Click" and the "Movement" are the two most important functions of the computer mouse.

To perform the experiments, the following items are included: 1) the subjects to perform the experiment: Three potential users were selected randomly from the lab. They were asked to perform several well-designed mouse-related actions. When performing the experiments, the subjects could see the effects of their motions on the screen. 2) A normal mouse and the Shoe-Mouse: Two kinds of mice were both tested for comparison. 3) A visualization module to calculate the efficiency of operation of the Shoe-Mouse in realtime. The following section presents the details.

1. *Experimental Result of the Click Motion Test.* The toe of the shoe contacts the ground with a force. By calculating the average value of this force, we can decide whether the user intends to produce a left click of the mouse. However, not all toe

motions can produce the corresponding click of the mouse. We did the following experiments to evaluate the performance of the click motion.

A circle with a 30-cm diameter was drawn manually on the screen. Each of the three volunteers wearing intelligent shoes performed a simple click of a toe to produce a point in this circle. If the click is successful, it will produce a new dark point inside the circle. Finally, the number of total points was computed to evaluate the efficiency of the proposed method. Figure 5.10 shows the result of the click test. From Figure 5.10, we can see that the successful rate of clicks of a person on the average is about 0.95, which is enough for the operation of a computer.

2. *Experimental Result of the Circular Motion Test.* In this experiment, the motion of the shoe as it was moving around a circular path was measured. We drew a circle as a test path, letting the users draw several identical circles in the same place as accurately as they could. Then we computed the deviation between the drawn circles and the original circle. If the deviation is bigger than the threshold, we considered this circle an unsuccessful circle. After that, we compared the perform-ance between the ordinary mouse and our proposed Shoe-Mouse. Although the performance of the ordinary mouse was a little better than the Shoe-Mouse, the Shoe-Mouse could perform ordinary tasks well enough that it can be used by a person to operate computers, especially for persons who have some difficulties in using their hands. Figure 5.11 shows the experiment and the performance. The left-most circle was drawn manually, and the middle circles were drawn by using the ordinary mouse, whereas the right-most circles were drawn using the Shoe-Mouse.

The above results show that the Shoe-Mouse can handle the operation of an ordin-ary mouse. It can detect the movement on the $x-y$ plane such that it can move the cursor and handle the click motion.

5.3.2 Intelligent Shoes for Pressure Measurement

5.3.2.1 *Motivation.* The processing of the pressure data beneath the foot provides a specialized form of human gait analysis. Information derived from plantar pressure data can give assistance in determining and managing the impaired symptoms associ-ated with a variety of musculoskeletal and neurologic disorders, which is of particular significance in such conditions as rheumatoid arthritis and diabetic neuropathy. Since

Person	Total-click	Sucessful-click
Person1	200	181
Person2	200	189
Person3	200	174
Average	200	181

Figure 5.10. Result of click test.

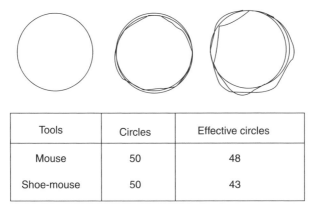

Tools	Circles	Effective circles
Mouse	50	48
Shoe-mouse	50	43

Figure 5.11. Performance of drawing a circle using the Shoe-Mouse.

the patients cannot sense the painful feeling of their soles correctly, the plantar pressure may be excessive. Applying the pressure higher than normal value will be an obstacle for blood reaching the tissues, which will result in ulceration, if prolonged. Based on the above reasons, different kinds of pressure measurement systems have been developed, either in the form of floor-mounted or in-shoe configuration.

Compared with floor-mounted systems, the in-shoe devices show more advantages. Subjects can wear the in-shoe device while walking in their normal gait without thinking about where the force platform is located. Multiple steps can be monitored by the in-shoe system but not with the force platform. Despite that the force platform system can provide both the shear and the vertical information of the ground reaction force, little loading information about the plantar surface with respect to the supporting surface is mentioned. In addition, with the development of wireless communication, the in-shoe device can be used not only in the clinic or laboratory, but also in the outdoor environment, which extends the usable locations for patients.

In the past decade, several researchers have developed in-shoe pressure measurement systems. In 1991, Hongsheng Zhu et al. developed a microprocessor-based data acquisition system to monitor the pressure distribution under seven bony prominences [10]. A wireless in-shoe force system was reported by Tracie L. Lawrence and Robert N. Schmidt in 1997 [11]. In this system, four thick-film force sensors were installed for each foot to estimate the total insole force and the center of pressure. The experiment results were compared with the Advanced Mechanical Technology, Inc., Watertown, MA (AMTI) force plate. The Pedar insole system (Novel, Munich, Germany) with 99 capacitance transducers for each insole is a commercially available system that is widely used in clinic sites and laboratories because of its repeatability and accuracy [12]. However, the limitations of this device include a heavy wireless and memory storage module, thick insole, and expensive price. The heavy weight of a wireless and memory storage module limits the period of gait trial and the applications in some violent activities, such as rapid running, dancing, jumping, and so on. Because the Pedar insole system uses the capacitance transducer, which is thicker (approximately 2 mm) in comparison with other types of sensors for in-shoe force measurement, it makes subjects feel a little uncomfortable when they wear their shoe together with this insole. The price is relatively expensive (approximately $31,000), which is unaffordable by most patients, even for some clinics.

In this part, a low-cost shoe-integrated system for measuring plantar pressure based on a support vector machine is developed. It can monitor not only the pressure distribution under the eight bony prominences but also the mean pressure beneath the entire foot. Ideal experimental results show that it is possible to use only eight FSRs to calculate the mean pressure, which used to be acquired by the device equipped with numerous sensors, such as the Pedar insole system. This intelligent system has the potential application for patients' gait analysis either in clinics or at home.

5.3.2.2   *Data Acquisition.*   The insole subsystem shown in Figure 5.12 is a flexible instrumented part for sensing the force parameters inside the shoe. Eight FSRs (Interlink Electronics, Santa Barbara, CA) are installed on one side of a thin insole made of plastic under subcutaneous bony prominences: 1–5 metatarsal heads, hallux (big toe), and the heel (which is divided into a posterior and inside portion). Considering the different sizes of bony prominences, we select two kinds of FSRs. Two FSR-402s (12.7 mm diameter active surface, 0.5 mm thick) are used in the first metatarsal head and hallux. Six FSR-400s (5 mm diameter, 0.4 mm thick) are placed under other positions.

FSR is a type of polymer thick film (PTF) device exhibiting a decrease in resistance when an increase in the force is applied to the active area [13]. In our circuit design, a voltage divider is used to measure the resistance change of the FSR in order to obtain the relationship between the applied force and the voltage.

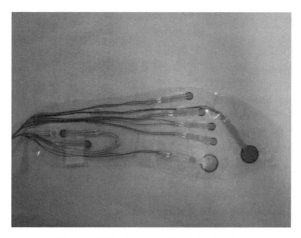

Figure 5.12. Photograph of the insole.

5.3.2.3 Sensor Calibration. To compensate the nonlinearity of FSR, each sensor needs to be calibrated after it has been located on the surface of the insole. A popular digital force gauge DPS-20 (Imada Incorporated, Northbrook, IL) is used to supply discrete force with the range from 0 kg to 10 kg for the type of FSR-400 and 0 kg to 20 kg for FSR-402. The digital outputs of force gauge are stored by PC via RS232. Then we can get the calibration result for each sensor according to the relationship between the applied force and the corresponding digital output of FSR. Experimental results demonstrate that a nine-order polynomial model provides a good fitting result. One calibration curve of FSR-402 under the first metatarsal head of right foot is displayed in Figure 5.13.

Equation 5.3 describes the relationship F between the FSR-402 output and the applied force in Newton:

$$F = -0.5801 \times x^9 + 6.426 \times x^8 - 28.15 \times x^7$$
$$+ 61.15 \times x^6 - 67.17 \times x^5 + 36.01 \times x^4 - 20.93$$
$$\times x^3 + 33.49 \times x^2 - 38.17 \times x^1 + 30.82 \qquad (5.3)$$

where x is the digital output of FSR-402 normalized by mean 369.7 and standard deviation 223.3.

After A/D transformation, the digital data of eight FSRs are packaged, which effectively decrease the transmission error and increase the sampling frequency to 50 Hz, which is adequate for the activity of walking [14]. Then in the part of the host computer, we obtain the force information applied for each sensor based on data reconstruction and calibration. Figure 5.14, displays the pressure waveform under each region as a function of time.

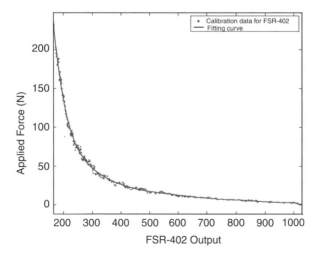

Figure 5.13. One FSR-402 calibration curve.

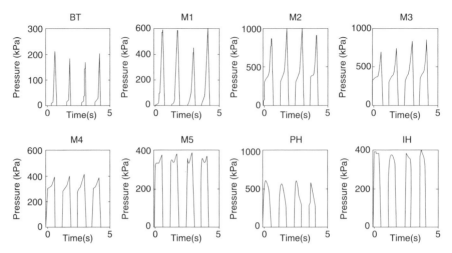

Figure 5.14. Pressure waveforms under eight regions of right foot during normal walking (BT = hallux, M1–M5 = 1–5 metatarsal head, PH = posterior heel, and IH = inside heel).

5.3.2.4 *Support Vector Regression.*

The feasibility of the support vector machine (SVM) in the extended application of regression problem has been proved, such as in the fields of electricity load forecasting [15], travel-time prediction [16], estimation of power system transient stability [17], and so on.

The basic idea of support vector regression (SVR) is to map a set of training points $\{(x_1, y_1), \ldots, (x_l, y_l)\}$, $(x_i \in X \subseteq R^n, y_i \in Y \subseteq R, l$ is the total number of training samples) from the input space X into a high-dimensional feature space via a nonlinear function ϕ in order that a hyperplane can be found approaching close to as many data points as possible in the feature space. SVR determines a function that can approximate future values accurately with the following linear form:

$$f(x) = (\omega \cdot \phi(x)) + b \tag{5.4}$$

Our goal is to find a function f to minimize the regularized risk function:

$$\begin{aligned} &\textit{Minimize} \quad \frac{1}{2}\|\omega\|^2 + C \sum_{i=1}^{l} L_\varepsilon(y_i, f(x_i)) \\ &\textit{Subject to} \\ &L_\varepsilon(y_i, f(x_i)) = \begin{cases} |y_i - f(x_i)| - \varepsilon, & |y_i - f(x_i)| \geq \varepsilon \\ 0, & \textit{otherwise} \end{cases} \end{aligned} \tag{5.5}$$

In Equation 5.3, minimizing the term $\|\omega\|^2$, which is called as the regularized term, will make the function as flat as possible. $\sum_{i=1}^{l} L_\varepsilon(y_i, f(x_i))$ representing the empirical risk is calculated by ε-insensitive loss, which is the most widely used cost function [18]. The constant C calculates the penalties to errors by determining

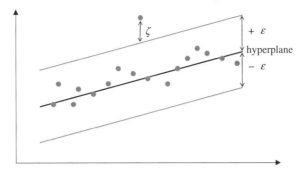

Figure 5.15. Graphical representation for SVR.

the trade-off between the empirical risk and the regularized term. The larger the value of C is, the more penalties are assigned to errors. ε denotes the tube size of the loss function. Both parameters C and ε are selected empirically by users.

By introducing positive slack variables ξ_i, ξ_i^* used to measure errors outside the ε tube, Equation 5.5 can be transformed into the primal objective function in Equation 5.6 to cope with otherwise infeasible constraints. This situation is shown in Figure 5.15.

$$
\begin{aligned}
\textit{Minimize} \quad & \frac{1}{2}\|\omega\|^2 + C\sum_{i=1}^{l}(\xi_i + \xi_i^*) \\
& y_i - \omega \cdot \phi(x_i) - b \leq \varepsilon + \xi_i \\
\textit{Subject to} \quad & \omega \cdot \phi(x_i) + b - y_i \leq \varepsilon + \xi_i^*, \quad i = 1,\ldots,l \\
& \xi_i^{(*)} \geq 0
\end{aligned}
\tag{5.6}
$$

We construct a Lagrange function L from Equation 5.4. Then from the partial derivative of L with respect to the primal variable ω, we can get the following equation:

$$
\omega = \sum_{i=1}^{l}(\alpha_i - \alpha_i^*)\phi(x_i)
\tag{5.7}
$$

By substituting Equations 5.7 into 5.4, the generic equation can be rewritten as

$$
\begin{aligned}
f(x) &= \sum_{i=1}^{l}(\alpha_i - \alpha_i^*)(\phi(x_i) \cdot \phi(x)) + b \\
&= \sum_{i=1}^{l}(\alpha_i - \alpha_i^*)K(x_i, x) + b
\end{aligned}
\tag{5.8}
$$

In Equation 5.8, α_i and α_i^* denote Lagrange multipliers satisfying positive constraints, $\alpha_i \times \alpha_i* = 0, \alpha_i \geq 0$. The dot product can be replaced with function $K(x_i, x)$ defined as the kernel function. The advantage of using the kernel function is that the dot product can be performed in high-dimensional feature space without having to know the nonlinear transformation $\phi(x)$ explicitly. Any function that satisfies Mercer's condition can be used as the kernel function. Radial basis function (RBF) kernel $K(x_i, x_j) = \exp(-\gamma \|x_i - x_j\|^2), \gamma > 0$ and polynomial kernel $K(x_i, x_j) = (x_i \cdot x_j + 1)^d$ are the commonly used kernel functions for the regression problem.

We can calculate the dual variable α_i, α_i^* by maximizing the dual problem:

$$
\begin{aligned}
Maximum \quad &-\frac{1}{2} \sum_{i=1}^{l} \sum_{j=1}^{l} (\alpha_i - \alpha_i^*)(\alpha_j - \alpha_j^*) K(x_i, x_j) \\
&-\varepsilon \sum_{i=1}^{l} (\alpha_i + \alpha_i^*) + \sum_{i=1}^{l} y_i(\alpha_i - \alpha_i^*) \\
Subject\ to \quad &\sum_{i=1}^{l} (\alpha_i - \alpha_i^*) = 0
\end{aligned}
\tag{5.9}
$$

Only several nonzero Lagrange multipliers α_i, α_i^* that fulfill the requirement can be used for the estimation of the regression line. As a remark, the sparsity of the support vector expansion is regarded because support vectors are usually a small subset of the training data points. Based on the Karush–Kuhn–Tucker (KKT) conditions [19], the constant offset b can be computed.

5.3.2.5 Experimental Results.

We use SVM to train the relationship between eight FSR data and the corresponding value of mean pressure gathered by the Novel Pedar insole system mentioned above. We process the SVR experiments running the SV package LIBSVM [20]. *C-SVM* proposed by Vapnik and *v-SVM* by Schölkopf et al. [21] are two kinds of SVM algorithms used in our experiments. As mentioned in Section 3, the performances of SVMs are sensitive to the selection of kernel functions and regularization parameters. After several experimentations, we chose the RBF kernel with the parameter γ equal to 0.5 as the kernel function for both *C-SVM* and *v-SVM* algorithms based on its positive performance in nonlinear mapping compared with other kernel functions.

In the *C-SVM* algorithm, we mainly pay attention to the influence of the user-defined parameter C that controls the balance between model complexity and the training error. The value range of C is from 0 to infinity; however, large C will result in an overfitting problem. We list the regression results of *C-SVM*, respectively, when C equals 1, 10, and 30 shown in Table 5.1. It can be found that larger C corresponds to the less mean-squared error (MSE), less the number of support vectors (SVs), as well as more iterations. When C equals 30, MSE can decrease at 0.865572 after 10649 iterations. Figure 5.16 displays the comparison of prediction results in *C-SVM* when different values of C are selected.

The *v-SVM* method is a new class of SVM that is closely related to the *C-SVM* procedure but with a different optimization risk function. In *v-SVM*, the optimal

Table 5.1. Regression Results of *C*-SVM with *C* = 1, 10, and 30

C	Iterations	SVs	MSE
1	1749	2462	1.18566
10	4694	2345	1.0854
30	10649	2306	0.835572

separating hyperplane is obtained by solving the following minimization problem:

$$\textit{Minimize} \quad \frac{1}{2}\|\omega\|^2 - \upsilon\rho + \frac{1}{l}\sum_{i=1}^{l}\xi_i$$

$$\textit{Subject to} \quad y_i(\omega \cdot \phi(x_i) + b) \geq \rho - \xi_i \tag{5.10}$$

$$\xi_i \geq 0, \quad i = 1, 2, \dots\dots\dots, l, \quad \rho \geq 0$$

where υ is the regularization parameter varying through [0, 1]. It limits the lower bound of the total support vectors and the upper bound number of the ones that lie

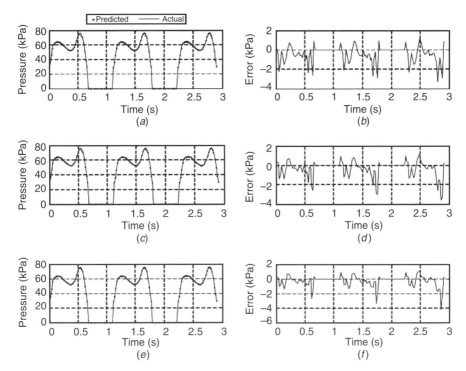

Figure 5.16. Comparisons of *C*-SVM predicted results using different parameters: (*a*) and (*b*) present the mean pressure and error when *C* = 1, MSE = 1.18566; (*c*) and (*d*) present the mean pressure and error when *C* = 10, MSE = 1.0854; (*e*) and (*f*) present the mean pressure and error when *C* = 30, MSE = 0.865572.

Table 5.2. Regression Results of ν-SVM with $\upsilon=0.1$, 0.5, and 0.8

ν	Iterations	SVs	MSE
0.1	11076	288	2.03745
0.5	103895	1371	0.929982
0.8	127855	2184	0.874932

on the wrong side of the hyperplane. The training of ν-SVM is more intuitive in the situation of selecting the parameter υ instead of C, which may lead to the overfitting problem in C-SVM. Table 5.2 shows the regression results of ν-SVM when υ equals 0.1, 0.5, and 0.8.

We compare the regression results of the two methods and find that the ν-SVM algorithm is less efficient than C-SVM on the problem at hand. When υ is selected as 0.8, the MSE of ν-SVM is a little more than the one of C-SVM when C equals 30. However, the value of iteration is 10 times more than the one of C-SVM. Figure 5.17 displays the prediction performance of ν-SVM with υ set to 0.1, 0.5, and 0.8.

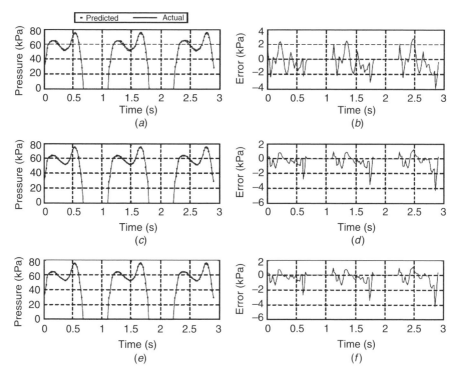

Figure 5.17. Comparisons of ν-SVM predicted results using different parameters: (a) and (b) present the mean pressure and error when $\upsilon = 0.1$, MSE = 2.03745; (c) and (d) present the mean pressure and error when $\upsilon = 0.5$, MSE = 0.929982; (e) and (f) present the mean pressure and error when $\upsilon = 0.8$, MSE = 0.874932.

5.3.3 Intelligent Shoes for Human Identification

5.3.3.1 Motivation. In this part, we focus on the research of identifying individuals through their walking patterns. Each person has a unique walking style. We can sometimes even recognize our friends by only looking at their walking styles from afar, or by listening to the sound patterns they make when they walk. The unique identity of a person can be identified by analyzing his/her fingerprints, voiceprints, and/or facial features. Similarly, analyzing the way people walk, their "step-prints," can also lead to the recognition of personal identities. Moreover, embedded force, inertial, and motion sensors in the Intelligent Shoe can offer important clues about the current activity of a user. Some previous work for human gait has mainly focused on foot parameter detection, such as temperature, humidity, heel-off time, gait velocity, and so on. As such, we propose to identify individuals by modeling their gait patterns with the integrated sensor data.

Hidden Markov models (HMMs) are doubly stochastic models and have been applied for a variety of stochastic signal processing. In speech recognition, where HMMs have found their widest application, human auditory signals are analyzed as speech patterns [22]. Transient sonar signals were classified with HMMs for ocean surveillance in Ref. 23. Radons et al. [24] analyzed a 30-electrode neuronal spike activity in a monkey's visual cortex with HMMs. Hannaford and Lee [25] classified task structure in teleoperation based on HMMs. We have previously proposed to use HMM for modeling and learning human skills. Nechyba and Xu [26] developed a validation method to compare the stochastic similarity between two dynamic, multidimensional trajectories using HMM analysis. Yang et al. applied HMMs to open-loop action skill learning [27] and to human gesture recognition [28].

The goal here is to design, build, calibrate, analyze, and use wearable intelligent shoes capable of measuring an unprecedented number of parameters relevant to gait, and to realize human identification based on gait modeling and similarity evaluation. The system is designed to collect data unobtrusively, and in any walking environment, over long periods of time. We treat gait as a human identity and a personal mark. Based on gait signal analysis, we can monitor human gait and identify the personal identity by his/her gait. A methodology based on modeling dynamic human gait and similarity evaluation with HMMs is proposed in this chapter. By achieving the modeling of human gait and physiological data, the identity of the wearer can be established through the gait pattern performance.

5.3.3.2 Data Acquisition. We select three single-axis gyroscope sensors and a three-axis accelerometer to detect the motion orientation of the foot. Three single-axis MEMS gyroscope sensors (ENJ-03J Murata) are mounted. As a miniature vibrating-read gyro, it uses piezoelectric material to sustain vibration, while taking advantage of Coriolis forces to measure angular rate. Each gyro sensor can detect the angular rate in one direction; thus, we can measure the yaw, roll, and pitch of shoe motion. Also, a three-axis MEMS accelerometer (MMA7260Q Freescale Semiconductor) is mounted, which can detect the acceleration motion of the shoes in three dimensions. The gyroscope sensors and accelerometer can detect three-

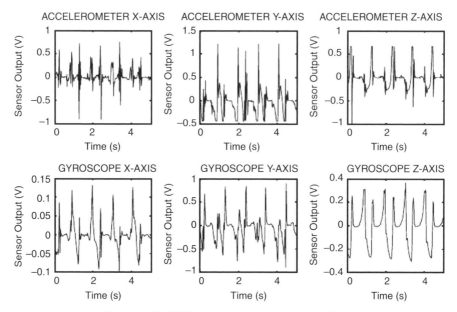

Figure 5.18. IMU waveforms during normal walking.

dimension rotation parameters and three-dimension acceleration parameters, which can be called an IMU. We use six dynamic signals of IMU as the input signals for gait modeling and similarity evaluation.

After A/D transformation, the digital data of IMU is packed, which effectively decreases the transmission error and increases the sampling frequency to 50 Hz, which is adequate for the activity of walking. Then as part of the host computer, we obtain the signal information applied for each sensor based on data reconstruction and calibration. Figure 5.18 displays the waveform of IMU as a function of time.

5.3.3.3 Hidden Markov Model. In general, gait data are dynamic, nonlinear, stochastic, time-varying, noisy, and multichannel. Data will not only vary among individuals but also with a single individual over time. Moreover, decisions in motion identification cannot be made on sensor readings at a particular instance, but rather the evolution of sensor data over time must be considered. Hence, we must select a modeling framework capable of dealing with these expected complexities in the data. Based on human gait modeling, we propose to compare the overall similarity among human walking patterns of several wearers using a probabilistic model that takes the information of human walking patterns into account.

We have developed an HMM-based similarity measure to compare the similarity among different human gait models. This similarity measure is based on HMMs, which are trainable statistical models with two appealing features: (1) No *a priori* assumptions are made about the statistical distribution of the data to be analyzed, and (2) a degree of sequential structure can be encoded by the HMMs.

The HMM is a collection of finite states connected by transitions. Each state is characterized by two sets of probabilities: a transition probability and a discrete output probability distribution or continuous output probability density function. In each state, an output symbol will be randomly produced. The HMM λ can be specified by three matrices.

$$\lambda = \{A, B, \pi\} \tag{5.11}$$

where A is the transition matrix that shows the probability of transitions among different states at any given state. B is the output probability matrix that shows the probability of producing different output symbols at any given state. π is the initial state probability distribution vector. For a given λ, it is capable of producing a series of output symbols that we call observation sequence O.

We define a stochastic similarity measure, based on discrete-output HMMs. Assume that we wish to compare observation sequences from two stochastic processes Γ_i and Γ_j. Let $O_i = O_i^{(k)}$, $k \in 1, 2, \ldots, n_i$, $i \in 1, 2$, denote the set of observation sequences of discrete symbols generated by process Γ_i. Each observation sequence is of length $T_i^{(k)}$, so that the total number of symbols in set O_i is given by,

$$T_i = \sum_{k=1}^{n_i} T_i^{(k)}, \quad i \in 1, 2 \tag{5.12}$$

The similarity measure σ between two observation sequences O_i, O_j is calculated using Equations 5.13 and 5.14, and the similarity distance measure ϕ is defined by Equation 5.15. We let P_{ij} denote the probability of the observation sequence O_i given by the model λ_j, normalized with respect to T_i, the length of the sequence. In practice, P_{ij} can be calculated using Equation 5.14 to prevent numerical underflow for long observation sequences.

$$\sigma(O_i, O_j) = \sqrt{\frac{P_{ji} P_{ij}}{P_{ii} P_{jj}}} \tag{5.13}$$

$$P_{ij} = P(O_i | \lambda_j)^{1/T_i} \tag{5.14}$$

$$\phi(O_i, O_j) = 10^{[P(O_i|\lambda_i) - P(O_i|\lambda_j)]/T_j} \tag{5.15}$$

Equations 5.16 to 5.18 show the properties of our similarity measure between two sequences O_i and O_j.

$$\sigma(O_i, O_j) = \sigma(O_j, O_i) \tag{5.16}$$

$$0 \le \sigma(O_i, O_j) \le 1 \tag{5.17}$$

$$\sigma(O_i, O_j) = 1, \quad if \, \lambda_i = \lambda_j \tag{5.18}$$

Given the definition of HMM, three basic problems of interest can be solved for real-world applications: the evaluation problem, the decoding problem, and the learning problem. The solutions to these three problems are the forward–backward algorithm, the Viterbi algorithm, and the Baum–Welch algorithm [29]. To develop the model state with the physical meaning of human gait, we use the Viterbi algorithm to segment the gait data sequences into states, and we study the properties of the spectral vectors that lead to the observations occurring in each state.

The human gait learning process can be connected with the similarity measure to iteratively improve a particular model. We envision that after a gait model with some stable pattern is trained, it can be stochastically perturbed in several possible directions, including structure, input representation, and choice of training data. The perturbation is parameterized into a perturbation vector $\Delta\theta_i$, and for the original action model $M(\theta_0)$ and the perturbed action model $M(\theta_0 + \Delta\theta_i)$ respectively, the similarity measure between the human gait data and the model-generated data is evaluated. The gradient of the similarity measure with respect to the variations in the model can be approximated and a better estimate of the motion model can be generated. Genetic approximation and simultaneously perturbed stochastic approximation (SPSA) are two techniques through which iterative model improvement can be achieved by gradient approximation. By applying the optimized action models, we believe that human gait can be identified more efficiently with lower cost and time.

5.3.3.4 *Human Gait Modeling and Similarity Evaluation.* In this section, we adopt HMM to account for gait modeling and similarity evaluation by modeling dynamic human gait as a separate HMM with its parameters learned from training data.

Although various types of HMMs have been proposed in the literature, we propose to apply the discrete HMM over its continuous and semicontinuous counterparts, because 1) it requires significantly less data to train reliably; 2) on average, it converges in many fewer iterations; 3) it is much less sensitive to initial random parameter settings during training; and 4) its orders of magnitude are less computationally expensive to train or evaluate. As sensor data are streamed from the intelligent shoes, it is first preprocessed through the same signal-conditioning algorithms as for the predefined gait models. The resulting observation sequence is then evaluated over the bank of gait models. Finally, the current short-term gait data are classified as that gait that has been evaluated to the highest observation probability.

We represent human gait as a time sequence. For each wearer, a separate N-state HMM will be built and the HMM training procedure will be used to optimally estimate HMM model parameters. Once the set of HMMs has been designed, optimized, and thoroughly studied, recognition of an unknown human gait is performed to score each individual gait model based on the given test observation sequence, and to select the wearer's identity whose model core is the highest matched.

1) Human Gait Modeling. We consider human gait data, including three-dimensional angular velocity and three-dimensional acceleration, as the observable stochastic process and the underlying stochastic process. Since the HMM has the ability to absorb the suboptimal characteristics within the model parameters,

human dynamic gait can be represented by transition possibilities and output possi-bilities, and using the same model we can "learn" human identification.

For each wearer, we want to design a separate N-state HMM. Let us consider the following procedure for the modeling of human gait: 1) Represent the human gait signal of a given wearer as a time sequence of coded spectral vectors, and have a train-ing sequence consisting of a number of repetitions of sequences of data indices of the individual gait. 2) Build individual wearer models. This task is performed by using the solution to train the HMM to optimally estimate model parameters for each model. To develop an understanding of the physical meaning of the model states, we use the Viterbi algorithm for the decoding problem to segment each training sequence into states, and then we study the properties of the spectral vectors that lead to the observations occurring in each state.

2) Similarity Evaluation. We use each set of observation sequences to train a corresponding HMM; this allows us to evaluate P_{ii} and P_{jj}. We then cross-evaluate each observation sequence on the other HMM to arrive at P_{ij} and P_{ji}. Finally we take the ratio of these probabilities and take the square root.

To apply this similarity measure toward comparing human dynamic gait, we need to convert real-valued sensor data to a sequence of discrete symbols. The human gait data are first normalized between $[-1,1]$. It is then segmentized into possibly over-lapping window frames. The Hamming window is applied to each frame to minimize spectral leakage caused by data windowing. The discrete Fourier transform is used to convert the real vectors to complex vectors. Using power spectral density estimation (Fourier), a feature matrix V is created. By applying Linde–Gray–Buzol (LGB) vector quantization algorithm, the feature vector V is converted to L discrete symbols such that the total distortion between the symbols and the quantized vectors can be minimized. The quantized sequence of human gait data will be used to train a six-state bakis HMM for similarity measure.

5.3.3.5 *Experimental Results.* In this experiment, we try to identify the person wearing the shoes by analyzing the gait performance. To estimate the gait perform-ance by the proposed system, we invited six human subjects to wear the intelligent shoes system. These subjects are HUANG, CHA, SHI, WANG, YE, and ZHONG. Each wearer walked on flat ground under normal speed. The sampling rate was set at 50 Hz based on the gait motion frequency. The system intercepted 1800 continuous data segments (sensor data in 36 seconds) for further modeling and training.

We first employed the fast Fourier transform (FFT) technique for preprocessing the six-dimensional time sequence $X(t)$. The Hamming window was first used with a width of 1.5 seconds (75 data segments). FFT analysis was then performed for every window, and the first three orders of FFT were kept for further precessing. After data pro-cessing, the original data matrix of size 1800 × 6 (gait data in 36 seconds) was changed to a matrix of size 75 × 18 each 1.5 seconds as one sequence for further training process. To model the human gait data $X(t) = \{A_x(t), A_y(t), A_z(t), G_x(t), G_y(t), G_z(t)\}$, where $A(t)$ represents the acceleration sensor data value of three dimensions and $G(t)$ represents the gyroscope sensor data value of three dimensions, a discrete six-state HMM was employed to encode human gait.

For the retrieved data segments, we employed the Baum–Welch algorithm to learn 975 of the segments for the HMMs and employed the forward–backward algorithm to evaluate the other 975 data segments. The training data were divided into 13 sequences, and each sequence contained 75 data segments. We introduced six-state left–right HMMs for modeling the gait of the six wearers. We initialized all HMMs parameters using a uniform segmentation of each training data sequence. Each sequence was split into six consecutive sections, where six was the number of states in the HMM. The vectors associated with each state were used to obtain initial parameters of the state-conditional distributions.

With these initial parameters, the forward–backward algorithm was run recursively on the training data. The Baum–Welch algorithm was used iteratively to re-estimate the parameters according to the forward and backward variables. Fifty iterations were run for the training processes. The forward algorithm was used for scoring each trajectory. Figure 5.19 shows all six HMMS forward scores of the results of the forward–backward algorithm, which is shown as the likelihood versus the learning iteration index.

The increase in score indicates the improvement of the model parameters. After the learning iterations, six HMMs were retrieved from the learning data samples. The HMMs λ_1 to λ_6 represent the human gait models from the six subjects who attended the test.

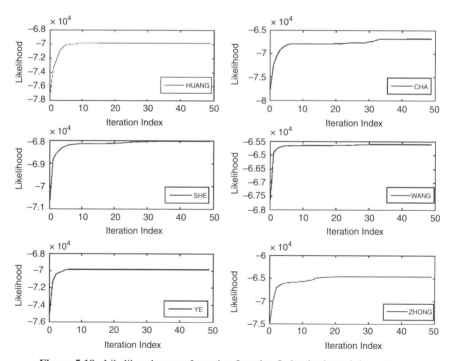

Figure 5.19. Likelihood versus Learning Iteration Index in the training process.

A) Model-to-Model Similarity Evaluation

To test and evaluate the trained models, we generated data vectors according to each given model. The data generated was the same as the training and testing set in size. Since each model represents individual gait, the HMMs λ_1 to λ_6, and the parameters N, M, A, B, and π were used as generators to give the observation sequence

$$O = O_1 O_2 \cdots O_M$$

where each observation O_k is one of the symbols from λ and M is the number of observations in the sequence, which is aforementioned as the simulated human gait data to test our models.

We used those output observations as the test input to all trained models. For each simulated observation O_{sim_i}, where i refers to the corresponding HMM λ_i, forward–backward algorithm was applied to all HMMs λ_j (in this experiment $1 \leq j \leq 6$). The following procedure can be used as a generator of observations:

1. Choose an initial state $q_1 = S_i$ according to the initial state distribution π.
2. Set $k = 1$.
3. Choose $O_k = v_k$ according to symbol probability distribution in state S_i, i.e., $b_i(k)$.
4. Transit to a new state $q_{k+1} = S_j$ according to the state transition probability distribution for state S_i, i.e., a_{ij}.
5. Set $k = k + 1$; return to step (3) if $k < M$; otherwise terminate the procedure.

The result P_{ij} is derived in the form of log-likelihood. In this experiment, where six human subjects were involved for testing, the complete log-likelihood is shown in Figure 5.20.

We also introduced similarity distance measure to reflect similarity evaluation. Deriving the log-likelihood $P(O_i|\lambda_i)$ and $P(O_i|\lambda_j)$ using the forward–backward algorithm, we obtained the similarity distance measure σ from λ_j to λ_i based on the definition of Equation 5.14. We applied the model-to-model similarity distance measure in Figure 5.21.

Likelihood	Model#1 HUANG	Model#2 CHA	Model#3 SHI	Model#4 WANG	Model#5 YE	Model#6 ZHONG
SS#1	−8748.9	−10015	−9678.6	−10520	−9274.3	−9602
SS#2	−8334.5	−7053.6	−8860.9	−7737.8	−8146.4	−8555.1
SS#3	−10073	−9988	−8412.9	−10034	−9298.8	−9346.7
SS#4	−7691.1	−7847.1	−8177.8	−6960.3	−7729.1	−8153.5
SS#5	−7213.6	−8181	−8088.9	−7939	−6663.2	−7825.3
SS#6	−8359.4	−8156.7	−8486.8	−8762.8	−8172.2	−7111.7

Figure 5.20. The likelihood of simulated sequences from HMMs to the corresponding HMMs.

Likelihood	Model#1 HUANG	Model#2 CHA	Model#3 SHI	Model#4 WANG	Model#5 YE	Model#6 ZHONG
SS#1	1	0.7166	0.783	0.6274	0.8709	0.7989
SS#2	0.6583	1	0.5543	0.7998	0.7	0.6125
SS#3	0.6349	0.6498	1	0.6417	0.7847	0.7745
SS#4	0.7852	0.7457	0.6685	1	0.7754	0.6739
SS#5	0.8268	0.5919	0.611	0.6435	1	0.6693
SS#6	0.6677	0.713	0.6407	0.5859	0.7094	1

Figure 5.21. The similarity distance measure of simulated sequences from HMMs to the corresponding HMMs.

To examine the results in Figure 5.20, in the ith row, the simulated test segment O_{sim_i} generated from HMM λ_i is as the input to all HMMs λ_j for the forward–backward algorithm, and the results P_{ij} can be seen that among all log-likelihood values for j HMMs, only when $i = j$, does log-likelihood achieve the maximum. In Figure 5.20, it can be clearly seen that the diagonal marked in gray achieves the maximum log-likelihood value among every single row, which can be explained as the data sequences generated from the trained models have much more similarity to the original data sequences from which the HMMs are trained than other models, or to say the HMMs are effectively to recognize similarity due to the data sequences, which is further applied to identify the wearers through their dynamic gait data sequences.

B) Human-to-Model Similarity Evaluation

To evaluate the HMMs for recognizing individual through the corresponding gait data sequences, we use the date segments, the same size as the training samples, from the testing set for individual identification. Like the procedure explained in Section 1, for each testing data sequence, we derive the log-likelihood values through the forward–backward algorithm to all trained HMMs and then the maximum log-likelihood is selected. Thus the individual identification is denoted as the corresponding HMM from which the maximum log-likelihood is derived.

We have 13 data sequences for each wearer in the test that were treated using the same preprocessing methods and have the same size. We selected two data sequences from each wearer for the identification test. The wearer's identities were counted each round, and the final results are shown in Figure 5.22. We also applied the human-to-model similarity distance measure in Figure 5.23.

C) Different HMM Structure

In the aforementioned discussion, the trained models are all based on the HMM structure of six states. We used different HMM structures and tested on the same testing data to find the best HMM structures for individual modeling method. In Figure 5.24, the testing results of average identification

Likelihood	Model#1 HUANG	Model#2 CHA	Model#3 SHI	Model#4 WANG	Model#5 YE	Model#6 ZHONG
GS#1	−7966.4	−8135.2	−8403.7	−8637.3	−8233.2	−8124.4
GS#1	−7950.7	−8484.1	−8374.6	−8697.8	−8031.4	−8048.1
GS#2	−8995.8	−7804.8	−9187.5	−8478.1	−9004.1	−9042.8
GS#2	−8436.7	−7268.1	−8346.8	−8188.4	−8485.8	−8280.7
GS#3	−9031.1	−9590.4	−8575.5	−9321.6	−8762.7	−9078.6
GS#3	−8645.8	−9412.2	−7932.5	−9428.7	−8553.9	−8731.1
GS#4	−8344.5	−8891.6	−8482	−7560.7	−8280.4	−8895.1
GS#4	−8376.6	−8640.4	−8333.4	−7684.2	−8260.6	−8799.6
GS#5	−8801.8	−9708.2	−8704.3	−10003	−8526.2	−9638.8
GS#5	−8503.2	−9188.9	−8626.5	−9608.7	−8327.4	−9012.4
GS#6	−8133.8	−8751.5	−8541	−8474.6	−8094.3	−7236.5
GS#6	−8165.8	−8568.8	−8454.3	−8268.4	−8231.8	−7409.4

Figure 5.22. The likelihood of gait sequence from individual wearer to the corresponding HMMs.

accuracy and the lowest identification accuracy of six wearers with different HMM structures are shown. We can see that the HMM with 9-state and 13-state can get the best performance than the other HMM structures. The average accuracy with a 9-state HMM structure is 91.03%, and the lowest accuracy is 61.54%; whereas the average accuracy with a 13-state HMM structure is 94.44%, and the lowest accuracy is 75.00%. Although 13-state corresponds to better performance than 9-state, the larger number of HMM states will increase model complexity, reduce model flexibility, and lead to overfitting in the training process. Additional explanation is required for the balance between the regularization term and the training errors. Thus, we

Likelihood	Model#1 HUANG	Model#2 CHA	Model#3 SHI	Model#4 WANG	Model#5 YE	Model#6 ZHONG
GS#1	1	0.9608	0.8893	0.8315	0.9341	0.9638
GS#1	1	0.8569	0.8845	0.8054	0.9769	0.9722
GS#2	0.7037	1	0.665	0.8198	0.702	0.694
GS#2	0.6906	1	0.7105	0.7471	0.6799	0.7256
GS#3	0.8879	0.7615	1	0.8185	0.951	0.8736
GS#3	0.813	0.6508	1	0.6477	0.835	0.7931
GS#4	0.7876	0.6668	0.7553	1	0.8032	0.6661
GS#4	0.8126	0.7509	0.8232	1	0.8414	0.7159
GS#5	0.9283	0.7267	0.953	0.6711	1	0.7405
GS#5	0.9526	0.788	0.9206	0.7017	1	0.8274
GS#6	0.7516	0.6175	0.6603	0.6744	0.7611	1
GS#6	0.7905	0.6975	0.7227	0.7657	0.7745	1

Figure 5.23. The similarity distance measure of gait sequences from individual wearer to the corresponding HMMs.

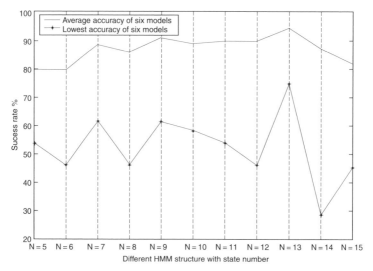

Figure 5.24. The identification success rate with gait modeling.

select the 9-state HMM structure in gait modeling and identification.

Furthermore, a total of 78 data segments from the six subjects were selected in the test with 9-state HMM structure, and their log-likelihoods values were derived by all trained HMMs from λ_1 to λ_6. The identification rate is shown in Figure 5.25.

5.4   CONCLUSION

In this chapter, we have built a shoe-based sensor-integrated platform that can gather much data inside a shoe. First, data are collected from different sensors installed in the shoe. Second, the data are computed and transmitted wirelessly to the host computer. Finally, the data are visualized on the screen. The platform is scalable and programmable. It can be used for many exciting applications such as gait recognition and human identification.

We introduced a novel input device called the Shoe-Mouse by applying the shoe platform as a user wearable interface. The Shoe-Mouse is especially designed for people who have difficulties in operating computers with their hands. As proved in the experiments, the Shoe-Mouse performs satisfactorily in motion control and can

Wearer	HUANG	CHA	SHI	WANG	YE	ZHONG	Total
Test sequent	13	13	13	13	13	13	**78**
Correct sequent	12	13	13	12	8	13	**71**
Identification rate	92.31%	100%	100%	92.31%	61.54%	100 %	**91.03%**

Figure 5.25. The identification success rate with 9-state HMM gait modeling.

partly replace the ordinary mouse. People who do a lot of work using computers can also benefit from this invention.

We presented a shoe-integrated plantar pressure measurement system based on the SVM to obtain the relationship between the data of eight force-sensing resistors fixed in insole and the mean pressure information gathered by the Novel Pedar insole system. Experimental results demonstrated that the regression model we built presents excellent performance of the in-shoe mean pressure prediction during the subject's normal walking. It has the potential application for aiding the patients with musculoskeletal and neurologic disorders in the development of normal gait.

We described an intelligent shoe system for collecting human dynamic gait performance. Using the proposed machine learning method HMMs, the individual wearer gait model was derived and we then demonstrated the procedure for recognizing different wearers through analyzing the corresponding models. Furthermore, we defined an HMM-based similarity measure that allows us to evaluate the resultant learning models. With the most likely performance criterion, it will help us to derive the similarity of individual behavior and corresponding model. The experimental results verify that the proposed method is valid and useful with a success human identification rate of about 91%.

REFERENCES

1. K. N. Tarchanidis and J. N. Lygouras, "Data glove with a force sensor," *IEEE Transactions on Instrumentation and Measurement*, Vol. 52, No. 3, pp. 984–989, June 2003.

2. R. E. Morley, E. J. Richter, J. W. Klaesner, K. S. Maluf, and M. J. Mueller, "In-shoe multi-sensory data acquisition system," *IEEE Transactions on Biomedical Engineering*, Vol. 48, No. 7, July 2001.

3. M. Skelly and H. Chizeck, "Real-time gait event detection for paraplegic FES walking," *IEEE Transactions on Systems and Rehabilitation Engineering*, Vol. 9, No. 1, pp. 59–68, Mar. 2001.

4. R. Williamson, and B. Andrews, "Gait event detection for FES using accelerometers and supervised machine learning," *IEEE Transactions on Rehabilitation Engineering*, Vol. 8, No.3, pp. 312–319, Sept. 2000.

5. L. Malone, C. Ellis-Hill, and I. Swain, "Using the Odstock dropped foot stimulator: user's and partner's perspectives," in *Proc. of 13th European Congress of Physical and Rehabilitation Medicine*, May 2002.

6. J. Paradiso, E. Hu, and K. Y. Hsiao, "The cybershoe: A wireless multisensor interface for a dancer's feet," in *Proc. of International Dance and Technology*, Tempe, AZ, 1999.

7. S. J. Morris and J. A. Paradiso, "A compact wearable sensor package for clinical gait monitoring," *Offspring*, Vol. 1, No. 1, pp. 7–15, January 31, 2003.

8. I. P. Pappas, T. Keller, and M. R. Popovic, "A novel gait phase detection system," in *Proc. of Workshop Automatisierungstechnische Verfahren für die Medizin*, Darmstadt, 1999.

9. A. Kale et al., "Identification of humans using gait," *IEEE Transactions on Image Processing*, Vol. 13, No. 9, pp. 1163–1173, Sep. 2004.

10. H. Zhu, G. F. Harris, J. J. Wertsch, W. J. Tompkins, and J. G. Webster, "A microprocessor-based data-acquisition system for measuring plantar pressures from ambulatory subjects," *IEEE Transactions on Biomedical Engineering*, Vol. 38, No. 7, pp. 710–714, July 1991.

11. T. L. Lawrence and R. N. Schmidt, "Wireless in-shoe force system," in *Proc. of the IEEE 19th Annual International Conference of Engineering in Medicine and Biology*, Chicago, IL, pp. 2238–2241, Nov. 1997.

12. H. Hurkmans, J. Bussmann, E. Benda, J. Verhaar, and H. Stam, "Accuracy and repeatability of the pedar Mobile system in long-term vertical force measurements," *Gait and Posture*, Vol. 23, No. 1, pp. 118–125, Jan. 2006.

13. www.micronavlink.com/

14. T. Mittlemeier and M. Morlock, "Pressure distribution measurements in gait analysis: dependency on measurement frequency," in *Proc. of 39th Annual Meeting of the Orthopaedic Research Society*, San Fransico, CA, 1993.

15. C.-C. Hsu, C.-H. Wu, S.-C. Chen, and K.-L. Peng, "Dynamically optimizing parameters in support vector regression: an application of electricity load forecasting," in *Proc. of the 39th Hawaii International Conference on System Sciences*, Vol. 2, No. 7, Jan. 2006.

16. C.-H. Wu, J.-M. Ho, and D. T. Lee, "Travel-time prediction with support vector regression," *IEEE Transactions on Intelligent Transportation Systems*, Vol. 5, No. 4, pp. 276–281, Dec. 2004.

17. D. H. Li and Y. J. Cao, "SOFM based support vector regression model for prediction and its application in power system transient stability forecasting," in *Proc. of IPEC 7th International Conference on Power Engineering*, pp. 1–6, Nov. 2005.

18. B. Schölkopf, C. J. C. Burges, and A. J. Smola, Eds., "Using support vector machines for time series prediction," in *Advances in Kernel Methods*, pp. 242–253, Cambridge, MA: MIT Press, 1999.

19. H. W. Kuhn and A. W. Tucker, "Nonlinear programming," in *Proc. of Berkeley Symp. Mathematical Statistics and Probabilistics*, pp. 481–492, 1951.

20. www.csie.ntu.edu.tw/~cjlin/libsvm/.

21. B. Schölkopf, A. Smola, R. C. Williamson, and P. L. Bartlett, "New support vector algorithms," *Neural Computation*, Vol. 12, pp. 1207–1245, 2000.

22. L. R. Rabiner, "A tutorial on hidden Markov models and selected applications in speech recognition," *Proc. IEEE*, Vol. 77, No. 2, pp. 257–286, 1989.

23. A. Kundu, G. C. Chen, and C. E. Persons, "Transient sonar signal classification using hidden Markov models and neural nets," *IEEE Transactions on Oceanic Engineering*, Vol. 19, No. 1, pp. 87–99, 1994.

24. G. Radons, J. D. Becker, B. Dulfer, and J. Kruger, "Analysis, classification and coding of multielectrode spike trains with hidden Markov models," *Biological Cybernetics*, Vol. 71, No. 4, pp. 359–373, 1994.

25. B. Hannaford and P. Lee, "Hidden Markov model analysis of force/torque information in telemanipulation," *International Journal of Robotics Research*, Vol. 10, No. 5, pp. 528–539, 1991.

26. M. Nechyba and Y. Xu, "Learning and transfer of human real-time control strategies," *Journal of Advanced Computational Intelligence*, Vol. 1, No. 2, pp. 137–154, 1997.

27. J. Yang, Y. Xu, and C. S. Chen, "Hidden Markov model approach to skill learning and its application to telerobotics," *IEEE Transactions on Robotics and Automation*, Vol. 10, No. 5, pp. 621–631, 1994.

28. J. Yang, Y. Xu, and C. S. Chen, "Human action learning via hidden Markov models," *IEEE Transactions on Systems, Man, and Cybernetics: Part A: Systems and Humans*, Vol. 27, No. 1, pp. 34–44, 1997.

29. L. R. Rabiner, "A tutorial on hidden Markov models and selected applications in speech recognition," *Proc. IEEE*, Vol. 77, No. 2, pp. 257–286, 1989.

CHAPTER 6

Fingertip Human–Computer Interface

6.1 INTRODUCTION

A computer is one of the most common devices in our lives now. Because of the development of mobile computing, pocket PCs or wearable computers allow computers and computer-related devices to merge human lives.

In this chapter, we discuss the advent in Micro ElectroMechanical System (MEMS) sensing and wireless technologies to develop a novel computer input system named the micro input devices system (MIDS, in short). It enables multifunctional input tasks and allows the overall shrinkage in size of the graphical user interface (GUI) interface devices (see Figure 6.1). The working principle lies on using the motion sensing technique to capture human motion and to convert the motion information to appropriate computer input commands. In terms of mobile computing, we envision MIDS to serve the functions of the current-day mouse, keyboard, and lightpen such that it will allow users to input text, draw graphical image, move curser, and perform drag-and-drop motion.

With the motion sensing ability of the MIDS, it can be further explored to measure the motion of robotic grasping manipulators and to control the grasping robot hands; hence, MIDS could replace the robotic gloves, which are currently used to interface with robot hands. The details will be shown in this chapter.

Conventional input devices, such as mouse and keyboard, and more advanced input devices, such as three-degree-of-freedom and six-degree-of-freedom controllers have relied on movement detectors, potentiometers, optical position detectors, liquid level tilt sensors, magnetic sensors, and button switches as inputs. Position is determined by reading the value of a movement detector and/or potentiometer attached to the input device. A potentiometer is simple and very inexpensive, and it is designed to work with standard computer interfaces. The potentiometer uses the desktop as a position reference. The biggest limitation of potentiometers is not performance, but the fact that they limit the motion and location of the user to a desktop location.

In addition, a traditional input device is a poor match to mobile devices [e.g., personal digital assistant (PDA)] and three-dimensional (3D) applications such as the playing of 3D, shooting, and racing games. A mobile computing user can use the

Intelligent Wearable Interfaces, by Yangsheng Xu, Wen J. Li, and Ka Keung C. Lee
Copyright © 2008 John Wiley & Sons, Inc.

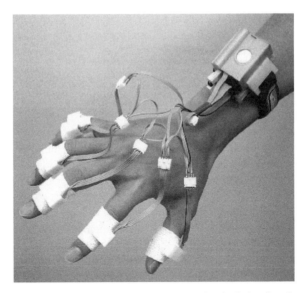

Figure 6.1. Wearable wireless MIDS prototype (this version includes five rings and one wrist watch).

simple stylus to control the PDA. A game player desiring a better experience can buy a separate steering wheel. What if it were possible to make a single controller that could simulate all of these?

Accelerometers are sensors that can measure acceleration. MEMS accelerometers can measure static gravity fields, allowing inclination measurement. Using gravity as a positional reference eliminates the need for working on the table and a connection to the desktop, i.e., available in "mid-air." Available tilt sensors are primarily of the electrolytic fluid variety. A glass or plastic vial is filled with a conductive fluid; electrodes in the fluid are used to sense a resistance, and as the device is tilted, the resistance changes. One problem with these sensors (which use gravity as an input) is their sensitivity to shock, vibration, and lateral acceleration. These signals, common to game playing, cause the liquid to slosh around in the vial producing a nonusable signal from the tilt sensor. These sensors typically have a response as poor as 0.5 to 1 Hz, about 50 times less than that required for a high-quality input device. Accelerometers used as inclinometers provide a better solution. New devices available today are just as accurate as conventional tilt sensors, but they have a response of better than 50 Hz and do not suffer the slosh phenomenon. They are also considerably smaller and lighter than liquid tilt sensors, allowing a reduction in the size and weight.

A generalized methodology for designing MIDS for different applications is covered in this chapter. MIDS combines MEMS accelerometers, wireless technology, and advanced algorithms to serve as a virtual mouse and keyboard that will allow users to input text, move cursor, perform drag-and-drop motion, and draw computer graphics (CG) on the desk or in mid-air. Therefore, this novel computer–human interface system allows the user to use only one input system to handle both graphical-based and text-based input interfaces. This work is also applied to the PDA for a motion-based controller successfully.

6.2 HARDWARE DESIGN

MIDS technology (MIDS-tech) is a method that makes use of motion data to generate operation commands for additional analysis. MIDS-tech includes the system design, software development, and motion-to-command algorithms so that it is a complete solution for handling motion analysis as well as for generating control-input commands.

A MIDS (a system made of one or more sensors and other peripheral subsystems such as Bluetooth wireless transmitters, power-storage units, etc.) can measure acceleration, velocity, and position of the fingertips, and thus, it can allow users to switch between different virtual input devices based on simple and customized finger motions to switch modes. In addition, a motion sensing device could also be developed based on MIDS-tech to analyze human motion in sports science, or as a monitoring system to keep in check equipment in an assembly line. MIDS could also be applied to robot-related applications such as in controlling the grasping motion of real/virtual robotic hands. Moreover, it could function also as a computer game controller to let players play a video game using their body motion.

MEMS sensors play a major role in MIDS in developing a functional system because of their low cost and miniaturized size. We used MEMS sensors to measure multidimensional motions of human body parts such as those at each finger and hand, followed by the wireless transmission of these motion data to the computer for input information process.

The MIDS has two versions: wired and wireless. Both of them consist of four main system units: 1) sensing unit including multiaxes MEMS sensors, 2) signal processing unit including microcontroller unit (MCU), 3) interface unit including the interface board connected to a PC or a laptop, and 4) driver interface program. For the wireless system, a wireless transmitter and receiver are connected to the MCU and to the interface board, respectively. The schematic diagram of MIDS is shown in Figure 6.2.

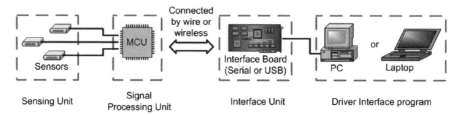

Figure 6.2. The MIDS schematic diagram including four main system units: 1) sensing unit, 2) signal processing unit, 3) interface unit, and 4) driver interface program.

The multiaxes MEMS sensors can be placed in MIDS rings or in any other form, which are worn on the fingers and are electrically connected with an MCU through wired or wireless transmission that acts as a communication link between the MCU and the PC. The MCU is used to analyze the sensor signals and to encode the signal for signal transmission. In the interface board, the analyzed signals are transmitted to the PC. It can use either a serial port or a USB port of the PC. Using the serial driver IC, MAX232 manufactured by Maxim Integrated Products Inc., Sunnyvale, CA [1], the received data are converted to RS232 format (computer-readable serial format) and then passed to the serial port of the PC. Using the USB driver IC (such as FT8U232BM manufactured by Future Technology Devices International Limited, Glasgow, United Kingdom [2]) and the USB port, the transmission rate can be increased and the PC can also supply the power through the USB port. The driver interface program is used to record the motion data and to convert the motion information to computer input command. It is a user-friendly interface, and it allows the users to change the parameters (e.g., the speed and sensitivity of the mouse cursor movement) for computer control.

6.2.1 MEMS Accelerator for Motion Detection

As mentioned, MEMS sensors play a major role in MIDS because they are low cost and have a miniaturized size. There are many types of MEMS sensors in the market such as accelerometer and gyroscope. Different sensors have different characteristics. MEMS sensors have been commonly used in the past. For example, Analog Devices, Norwood, MA [3], Memsic Inc., North Andover, MA [4], and STMicroelectronics, Geneva, Switzerland [5] designed and manufactured their own MEMS sensors.

6.2.1.1 Accelerometer. Accelerometer is used to measure the acceleration of a moving object. It is commonly used in automotive, automation machines, measuring instruments, and monitoring systems. For example, acceleration signals can be integrated with velocity and can be doubly integrated to form displacement. Accelerometer can detect the motion, vibration, and shock of machines and automobiles, and they can measure gravity such that it can be used to determine the orientation of the object such as tilt and inclination.

In this work, MEMS accelerometers manufactured by Analog Devices Inc. are used. These sensors are produced by surface micromachining. Surface micromachining is a technique for building MEMS structures in silicon. Combined with onboard signal conditioning circuits, complete micro electromechanical systems can be built on a small piece of silicon such that the size of sensors can be in micro-scale. Therefore, surface micromachining allows the consistent and repeatable production of large quantities of devices at low cost. It is the reason for choosing these MEMS sensors.

There are three types of MEMS accelerometer: single axis, dual-axis, and three-axis (see Figure 6.3). These accelerometers can be used to measure the acceleration in Cartesian space (xyz directions).

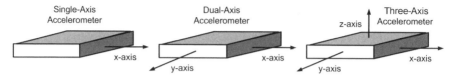

Figure 6.3. Three types of accelerometer: single axis, dual-axis, and three-axis.

The MEMS sensors are capable of measuring acceleration with a full-scale range from below ± 1.0 g to above ± 100 g (Note that 1.0 g $= 9.81$ ms^{-2}). The sensors can measure both dynamic acceleration, such as vibration and shock, and static acceleration such as gravity. The sensors employ the principle of relating the capacitance variation between the polysilicon comb-drives to acceleration (as illustrated in Figure 6.4).

When movement is applied to a sensor, the proof mass is moved such that the capacitances between the two fixed outer plates (C1 and C2) are changed. The acceleration can then be determined by the ratio of the capacitances. With different onboard signal conditioning circuits, the sensors can provide analog or digital output voltages. The analog output voltages are available in absolute voltage (which is independent of supply voltage) and ratiometric voltage (which is proportional to supply voltage). The digital output voltages are signals with a duty cycle (which is the ratio of pulse width to period and it is in the form of a pulse-width module, PWM) [6].

As an example, a capacitive accelerometer can be made by using silicon MEMS etching techniques. A model is shown in Figure 6.5, where m is the proof mass, c is the damping coefficient, k is spring constant, δ_o is the gap distance, ξ is the input motion and x_o is the output motion. The input motion will change the gap distance δ (with the initial gap distance δ_o), thereby changing the capacitance between two electrodes. Hence, the acceleration can be measured by monitoring the capacitance.

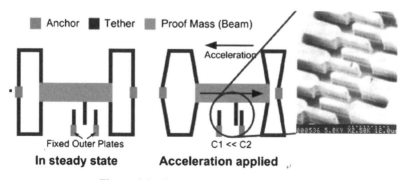

Figure 6.4. Illustration of sensor operation.

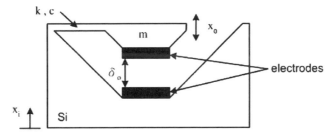

Figure 6.5. Model of a MEMS capacitive accelerometer.

The capacitance between the electrodes is

$$C = \frac{\varepsilon_o A}{\delta_o} \tag{6.1}$$

Even though capacitive sensing technique is matured now, it still has some problems when the mechanical sensors are miniaturized: 1) matching air-gap and CMOS capacitors, 2) thermal drift, 3) variation in driving voltage, and 4) variation of δ_o of air gap. Except for the thermal drift problem, we can gain an insight into these problems by doing the following analyses.

Assume that a simple capacitive circuit has a capacitor that will change its capacitance due to a mechanical input, the circuit analysis diagram for capacitive accelerometer modeling is shown in Figure 6.6.

Then, from the circuit analysis:

$$V_o = V_{AC} \frac{Z_{\text{ref}}}{Z_S + Z_{\text{ref}}}, \quad \text{where} \quad Z_{\text{ref}} = \frac{1}{j\omega C_{\text{ref}}} \quad \text{and} \quad Z_S = \frac{1}{j\omega C_S} \tag{6.2}$$

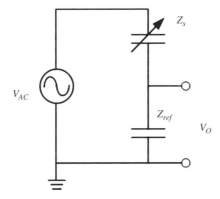

Figure 6.6. Circuit analysis for capacitive accelerometer modeling.

Hence,

$$V_o = V_{AC} \frac{C_S}{C_{ref} + C_S} \tag{6.3}$$

Let,

$$C_S = \frac{\varepsilon_o A}{\delta_o + d\delta} \tag{6.4}$$

where δ is the gap distance between the Z electrodes in the sensing capacitor of any given time.

For the initial capacitance,

$$C_{So} = \frac{\varepsilon_o A}{\varepsilon_o} = C_{ref} \tag{6.5}$$

Thus,

$$V_o = V_{AC} \frac{\dfrac{\varepsilon_o A}{\delta_o + d\delta}}{\dfrac{\delta_o A}{\delta_o} + \dfrac{\varepsilon_o A}{\delta_o + d\delta}} = \frac{V_{AC}}{2} \left(\frac{1}{1 + \dfrac{d\delta}{2\delta_o}} \right) \tag{6.6}$$

Then, from the circuit analysis,

$$V_o = V_{AC} \frac{C_1}{C_1 + C_2} \tag{6.7}$$

Let the capacitances between the electrodes be

$$C_1 = \frac{\varepsilon_o A}{\delta_o - d\delta} \quad \text{and} \quad C_2 = \frac{\varepsilon_o A}{\delta_o + d\delta} \tag{6.8}$$

Then,

$$\frac{C_1}{C_1 + C_2} = \frac{\delta_o + d\delta}{2\delta_o} \tag{6.9}$$

Thus,

$$V_o = V_{AC} \frac{\delta_o + d\delta}{2\delta_o} \tag{6.10}$$

Hence, the output voltage of a MEMS sensor can be directly related to the displacement of a proofmass as shown in the above equation.

The ADXL202E is a MEMS-based dual-axis accelerator manufactured by Analog Devices Inc. For each axis, an output circuit converts the analog signal to a duty cycle modulated (DCM) digital signal that can be decoded with a counter/timer port on a microprocessor. The ADXL202E is capable of measuring both positive and negative accelerations (the range is ± 2 g where g means $9.8\,\text{ms}^{-2}$). The accelerometer can also measure static acceleration forces such as gravity, which allows it to be used as a tilt sensor.

The ADXL202E's digital output is a duty cycle modulator. Acceleration is proportional to the ratio $T1/T2$. The nominal output of the ADXL202E is [3]:

1. 0g= 50% duty cycle
2. Scale factor is 12.5% duty cycle change per g

These nominal values are affected by the initial tolerance of the device including 0g offset error and sensitivity error. It is necessary to perform a calibration. The period of duty cycle, $T2$, does not have to be measured for every measurement cycle. It need only be updated to account for changes due to temperature (a relatively slow process). Since the $T2$ time period is shared by both X and Y channels, it is necessary only to measure it on one channel of the ADXL202E. The acceleration value A (in terms of g) is

$$A(g) = (T1/T2 - 0.5)/12.5\% \tag{6.11}$$

With the acceleration information in x- and y-directions, the pitch and roll angle can be determined as follows:

$$\theta_{\text{pitch}} = \sin^{-1}(A_y(g)/g) \tag{6.12}$$

$$\theta_{\text{roll}} = \sin^{-1}(A_x(g)/g) \tag{6.13}$$

ADXL202E can be used to measure a full $360°$ of orientation through gravity (see Figure 6.7). When one axis output is reading a maximum change in output per degree, the other is at its minimum ($90°$ perpendicular to each other).

To find the tilt angle α in $360°$ scale, it can be defined as four phases. It can be obtained by the value of the acceleration in y- and x-directions according to the phases. The detailed formulas are shown in Table 6.1.

6.2.2 Signal Processing and Analysis

Many MCUs can handle a fast-speed DCM signal. The Atmega8515 manufactured by Atmel Corporation [7] is selected in our work. In the signal processing unit of MIDS, a programmable MCU is used to count the duty cycle of the ADXL202E such that the acceleration values can be determined. The illustration of the data acquisition of the MCU for DCM signal from ADXL202E is shown in Figure 6.8.

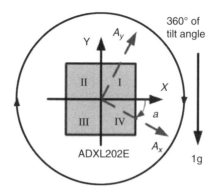

Figure 6.7. Using x- and y-direction output to measure 360° of tilt.

Table 6.1. Detailed Formulas for Calculating the Tilt Angle α in 360° plane

Phase	Range of Y	Range of X	α
I	$0 \leq Y \leq 1$	$-1 \leq X \leq 0$	$\alpha = \sim^{-1}(A_y(g)/g)$
II	$0 \leq Y \leq 1$	$0 \leq X \leq 1$	$\alpha = 90° + \sim^{-1}(A_x(g)/g)$
III	$-1 \leq Y \leq 0$	$0 \leq X \leq 1$	$\alpha = 180° + \sim^{-1}(A_y(g)/g)$
IV	$-1 \leq Y \leq 0$	$-1 \leq X \leq 0$	$\alpha = 270° + \sim^{-1}(A_x(g)/g)$

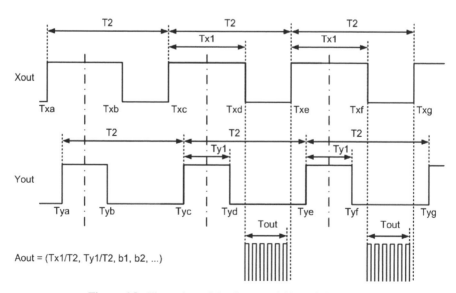

Figure 6.8. Illustration of the data acquisition of the MCU.

In Figure 6.8, Xout and Yout are DCM signals for x-direction and y-direction from ADXL202E. The points Txa, Txc, Txe, and Txg are the raising edge of Xout, whereas the points Txb, Txd, and Txf are the falling edge of Xout. Similarly, the points Tya, Tyc, Tye, and Tyg are the raising edge of Yout whereas the points Tyb, Tyd, and Tyf are the falling edge of Yout. Since the T2 for Xout and Yout are the same, the period of duty cycle can be calculated from either Xout or Yout to get the value of T2.

At the beginning of the program, the first raising edge will be recorded. Assume Xout is faster than the Yout (see Figure 6.8); Txa will be recorded in order to determine the start of the T2. When the second raising edge of Xout appeared, T2 can be determined by (Txc − Txa) and Txc will then be recorded in order to determine the start of Tx1.

At this stage, the MCU will monitor both Xout and Yout to record the raising or falling edges. Thus, Tyc from Yout will then be recorded in order to determine the start of Ty1. When a falling edge appeared in Yout, Tyd will then be recorded to determine Ty1. After that, a falling edge appeared in Xout so the Txd will be recorded to determine Tx1. Now, T2, Tx1, and Ty1 have been determined. For the ADXL202E, the acceleration value Aout for x- and y-directions can be calculated and represented by 8-bit signals.

On the other hand, the button press action is also checked. Through the Universal Synchronous and Asynchronous serial Receiver and Transmitter (USART) from the MCU, the Aout signals for x- and y-directions as well as the button status (b1, b2, etc.) will then be transmitted in serial format within the time T2 − T1 (i.e., Tout <T2 − T1). Then the Tx1 or Ty1 of the next duty cycle will then be determined continuously.

Note that Tout must be less than T2−Tx1; otherwise, there will be a data lost problem. The details to prevent this problem will be discussed in the coming section on the different applications. T2 will be updated every second (the threshold is 150, i.e., 150 samples per second) so that the error due to temperature change will be minimized. The program flow of the signal processing unit is shown in Figure 6.9.

6.2.2.1 *Noise Filtering and Signal Processing.*

For some applications such as human–machine interface and computer input device, the high-frequency noise signal (which sometimes is from the human hand shaking) should be eliminated. A low-passed filter is adopted to eliminate such high-frequency noise. When the driver interface receives the sensor data and obtains the acceleration value, the acceleration signals are first passed through a first-order Butterworth filter before converting into the control commands. The transfer function of the first-order low-pass Butterworth filter is

$$H(s) = \frac{\Omega}{s + \Omega_c}, \text{ where } \Omega = \tan\frac{\pi f_c}{f_s} \tag{6.14}$$

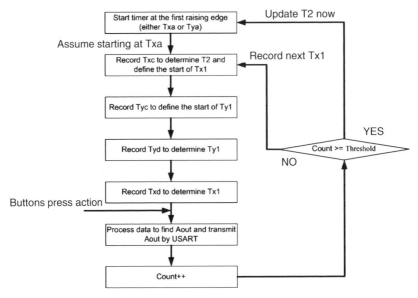

Figure 6.9. Program flow of signal processing unit.

Note that f_c is the cutoff frequency (Hz) and f_s is the sampling frequency (Hz).

$$s = 2f_s \frac{1 - z^{-1}}{1 + z^{-1}} \tag{6.15}$$

With bilinear transformation, the digitalized expression of the output in time domain is

$$y(n) = \frac{\Omega_c}{2f_s + \Omega_c} u(n) + \frac{\Omega_c}{2f_s + \Omega_c} u(n-1) + \frac{2f_s - \Omega_c}{2f_s + \Omega_c} y(n-1) \tag{6.16}$$

Note that the cutoff frequency is selected to be $= 25\,\mathrm{Hz}$ for human motion application, which is double that of the maximum frequency of human hand motion ($\sim 10\,\mathrm{Hz}$). For other applications such as machine monitoring, the cutoff frequency should be based on the specific requirement.

6.2.3 Radio-Frequency (RF) Wireless System

Many RF wireless solutions are available in the market. A pair of LC series transmitter and receiver modules manufactured by LINX technology [8] is selected. The LC series wireless module is capable of transferring serial data at distances in excess of 300 feet. No external RF components, except an antenna, are required. Moreover, it is low cost, has ultra-low-power consumption, is a compact surface-mount package, has fast enough data rates to 5000 bps, and has a direct serial interface that supports serial data transmission.

6.2.4 System Evaluation

Experiments were performed to demonstrate motion detection in 3D space using a simple version of MIDS. This simple version of MIDS includes two 3D motion sensing rings and one MIDS watch with signal processing unit and wireless transmitter. An illustration of the components of the MIDS ring is shown in Figure 6.10. Two dual-axis MEMS sensors are mounted as shown. Sensor A is placed at the top of the ring horizontally to measure fingertip accelerations in the x- and y-directions. Sensor B is placed at the side vertically to detect accelerations in the y- and z-directions. Therefore, sensor A can detect the plane motion of the fingertip and sensor B can detect the fingertip angle (relative to rotation about the mid-joint of a finger) and the vertical movement. The prototype of this simple MIDS ring is shown in Figure 6.11. On the other hand, the receiver was connected to the PC and the data were collected for analysis.

Two experiments were performed: 1) measurement of the fingertip motion in z-axis for click motion detection, and 2) measurement of the fingertip movement on x–y plane to determine the trajectories of the movement as sensed by the motion sensors. The experimental setup is shown in Figure 6.12. The data acquisition schematic is shown in Figure 6.13.

6.2.4.1 Experimental Results of Click Motion Test. Measurement for click motions in z-direction is shown in Figure 6.14, which basically shows the up and down motions of the fingertip. Three cases are distinctive in the data plot: (a) one peak, (b) two peaks, and (c) three peaks. As indicated in the figure, the time duration for each sampled motion is about 1 second. Regions (a), (b), and (c) shown in the figure correspond to single-click, double-click, and triple-click, respectively. These results show that the accelerometer could detect the click motions (even for triple-click) within a short time (about 1 second). The sampling frequency used for this

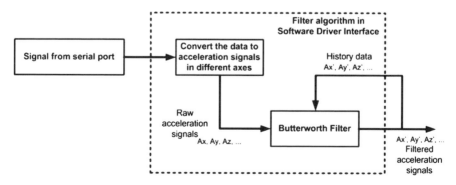

Figure 6.10. Schematic of MIDS ring.

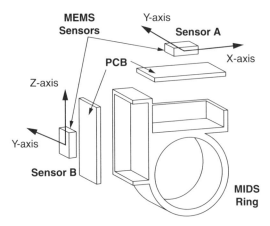

Figure 6.11. Prototype of simple MIDS rings.

experiment is about 31 Hz. The response time of the sensors are fast enough to measure the fingertip motions based on these experimental observations.

6.2.4.2 *Experimental Results of Circular Motion Test.* The motion of a fingertip as it traverses around a circular path was also measured. (The circular path was imagined by the experimental subject; no reference circular path was placed on a table for the subject to follow.) The time-sequence data of the two sensors on the fingertip is shown in Figure 6.15, and the corresponding displace plot in the x–y plane is shown in Figure 6.16, which indicates that the MIDS could potentially be used to perform the functions of moving a cursor on the computer screen.

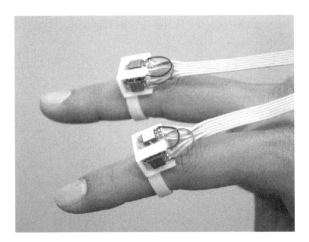

Figure 6.12. Experimental setup for motion tests.

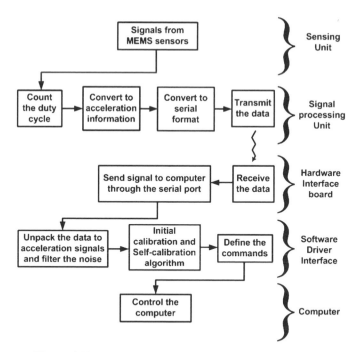

Figure 6.13. Data acquisition schematic of the simple MIDS.

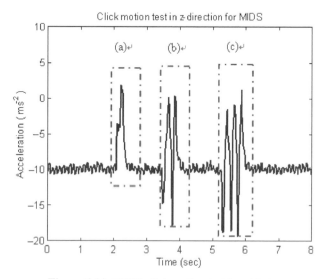

Figure 6.14. MIDS click motion test in mid-air.

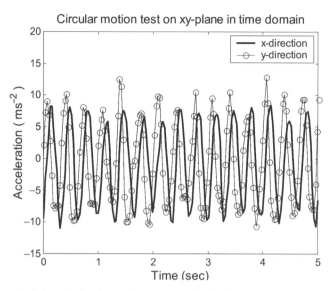

Figure 6.15. MIDS circular motion test on a desk (time sequence acceleration).

With the above results, we have shown that the MIDS could potentially handle the operations of existing common input devices. For the mouse, MIDS could detect the movement on x–y plane such that it could move a cursor and handle the click motion; for the keyboard, MIDS could detect the finger movement (as a finger moves from one key to another) and the key-press action (as the finger types a key); for the

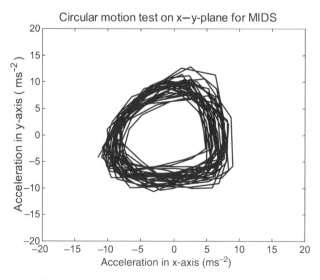

Figure 6.16. MIDS circular motion test on a desk.

light pen, MIDS could capture the trajectory of the finger such that it can draw a desired curve in either a fixed two-dimensional (2D) plane or in 3D space.

6.3 SPECIFIC APPLICATIONS

Motion is a general term describing the act or process of changing position and place of a dynamic object. Motion is really a very important part in our life. For sports, many athletes use equipment to evaluate their motion during exercises. For industry, many engineers use motion sensing systems to monitor machines in the assembly lines.

As mentioned, MIDS can measure many types of motion. Using MIDS technology, different motions sensing systems can be developed. In this section, motion sensing devices (MSDs) developed for human–robot interaction, computer-mouse on a fingertip, computer game control, and PDA interation are presented.

6.3.1 Human–Robotic–Hand Interaction Using MIDS

Nowadays, fingered robots are becoming more ubiquitous, e.g., in the manufacturing industry, robot hands are now used to handle many duplicated tasks such as grasping in product assembly line. Human-like robots such as Honda's humanoid robot: ASIMO [9] and Sony's SDR-4X [10] are commercially available in the market. These robots can be classified into two different types: those controlled by the CPU automatically and those manipulated by operators. For automated robot hands, MSDs are essential for measuring the motion of the robot hands for quality control and system evaluation. For manual controlled robot hands, a user-friendly input device can be useful to the operator in controlling manipulators such as grasping robotic hands. In this case, a multifunctional device is needed to handle both the motion sensing function and the control-input function. In this section, we will show that MIDS will enable many new capabilities in terms of robotic sensing and control, including a multifunctional device for sensing and controlling robotic hands.

In the robotics area, studies in grasping using robotic hands have been extensive in the past decade [11–13]. Some researchers have focused on theoretical analyses. For example, Ding et al. proposed an algorithm for computing a form-closure grasp on a 3D polyhedral object [11]. The computational cost of this algorithm is less dependent on the complexity of the object surface such that it can help the robotic hand to grasp objects with many contact points. Also, Hasegawa et al. proposed a method to generate the grasping point and the regrasping phase [12]. Moreover, Cheah et al. proposed a method to solve a task-space feedback control problem of multifingered robotic hands with uncertain Jacobian matrices [13].

In this section, we try to further explore MIDS to ultimately replace the robotic gloves that are used today to interface with virtual and real robotic hands. MIDS not only can be used to develop or evaluate control algorithms by measuring the motion of robotic hands, but it may also act as a human–robot interface that allows users to control the robotic hands by their own finger motion.

6.3.1.1 Human Finger Motion Analysis. To design a user-friendly human–robot interface and an effective method for robotic hand control, it is very important to study the characteristics of the human hand and finger motion. In this study, the maximum motion frequency (or the working frequency range) of each human finger has been determined. The results of this study will also serve as a valuable reference for the characteristics of human finger motions, which can be used to develop more advanced MIDS applications such as a virtual keyboard.

A typical computer user is selected, who is good at typing, to perform up and down motion (like click motion) as fast as possible for each finger for 100 cycles (the user worn a MIDS ring sensor in each finger). The experimental results for each finger in terms of acceleration power spectrum density (PSD) are shown in Figure 6.17.

Five graphs are shown in Figure 6.17, respectively, for the thumb finger, index finger, middle finger, ring finger, and little finger. As shown, the maximum acceleration PSD of the thumb, index and middle fingers are higher than the ring and little fingers and the peaks for these three fingers are around 6–8 Hz. Moreover, the acceleration PSD levels of the index and middle fingers among the whole frequency range are comparatively higher than the other three fingers. Note that the scale y-axis for the

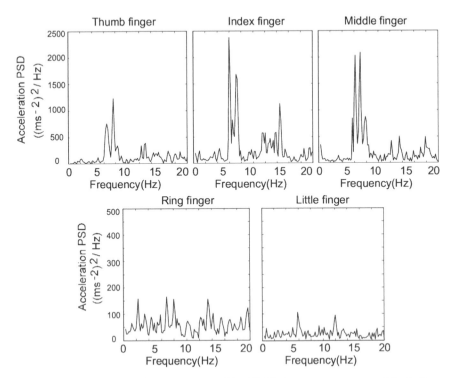

Figure 6.17. Acceleration PSD data measured by MIDS rings worn on a test subject's finger as the subject typed as fast as he can for 100 times using each finger.

graphs in Figure 6.17 of the ring and little finger are set to $0-500 \ (\text{ms}^{-2})^2 \ / \ \text{Hz}$. This result shows that the moving ability of the index and middle finger are higher than the thumb, ring, and little fingers. Since this experiment was done by performing 100 cycles of up and down motion using the best reflex of the experimental subject, the working frequency range for the five fingers can be determined by checking the peaks of the acceleration PSD, which is less than 20 Hz.

Since the human fingers have only $\sim$20 Hz reflex motion, it is reasonable to filter out any noise signal above this frequency for the finger-to-robot interface application. Therefore, to eliminate the high-frequency noise signal, the sensor data were filtered by a first order Butterworth filter. (The computation was performed on a computer that received the MIDS data wirelessly, processed the data, and then sent it to a robotic hand controller). According to the result of Figure 6.17 the cutoff frequency should be selected by $f_c = 20$ Hz, which is higher than the maximum moving frequencies of five fingers and the sampling frequency $f_c = 42.37$ Hz which is higher than twice the cutoff frequency.

6.3.1.2 Experimental Setup. Two experiments were performed to demonstrate the grasping motion control of a five-finger robotic hand wirelessly by using MIDS. The manipulator used in the experiment is a five-finger robotic hand system in the Robot Control Laboratory of The Chinese University of Hong Kong. Each robotic finger made by Yaskawa has three revolute joints driven by AC motors through a harmonic drive of a 80:1 reduction ratio. The robotic hand is controlled by a distributed DSP C40's system. The coordinate definition of the robotic hand system is shown in Figure 6.18.

The general process for using MIDS to control robotic fingers is described below. The grasping motion of the operator is first detected by the MIDS rings. The acceleration signal is then passed to the microprocessor of the MIDS controller for signal encoding. After that, the packed signal is transmitted to the receiver through the wireless transmitter inside the MIDS controller. Once the wireless receiver receives the signal, the interface board decodes the received signal and then passes the acceleration signals to the PC through the serial port. A PC, which is connected to the C40, is used to transmit the control commands to the robotic hand system.

The MIDS rings were positioned on each finger of the operator. The configuration of the MIDS and the five-finger robotic hand are shown in Figure 6.19. The configuration for the robotic hand is $-90°$ defined such that the body coordinate frame of the robotic finger is rotated about the x-axis of the world coordinate frame (i.e., the x$-$y plane is orthogonal to the horizontal plane of the world coordinate frame, and the z-axis is orthogonal to the vertical axis).

6.3.1.3 Control Algorithm. In the experiments, all fingers rotated about the x-axis (see Figure 6.19). The tilt angle of each operator's finger α_i, is measured by MIDS where $i = 1, \ldots, 5$, respectively. We have developed an algorithm to control the rotational angle of each robotic finger θ_i, which was tracked to the tilt angles α_i. We used a PD control tracking method to control the rotational angles of

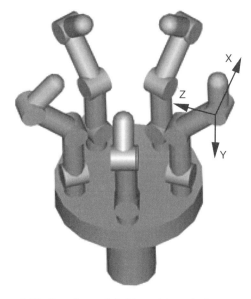

Figure 6.18. Coordinate definition of the robotic hand system.

robotic fingers. We define the joint angle errors e_i, as the differences between the human finger angles α_i and the robotic finger angles θ_i such that:

$$e_i = \alpha_i - \theta_i \tag{6.17}$$

Then the control law of PD controller can be defined as

$$\tau_{i,j} = \tau_{i-1} + P_i(e_{i,j} - e_{i,j-1}) + D_i(e_i - 2e_{i,j-1} + e_{i,j-2}) \tag{6.18}$$

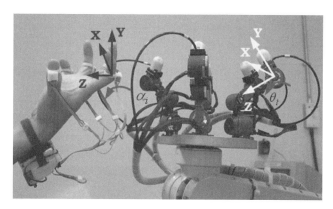

Figure 6.19. Configuration for robotic hand and MIDS.

where τ_i are the control torques for the five fingers, $i = 1, \ldots, 5$; j is the sequential variable; $P_i = 20$ are the position gains for the five fingers, $i = 1, \ldots, 5$; and $D_i = 100$ are the velocity gains for the five fingers, $i = 1, \ldots, 5$.

The control torques τ_i are the input forces for the joint motors of the five robotic fingers. The joint torques are controlled to change the joint angle of each robotic finger. For the arm control of the robotic hand, the measured joint angles were directly sent to the robotic hand system. The robotic arm can then directly move to that angle without an extra control algorithm.

6.3.1.4 Experimental Results. Two experiments were performed to demonstrate the controllability of moving and grasping motions of the robotic hand manipulator. In the first experiment, the operator tried to perform the grasping motion by opening the hand and closing the hand periodically. The experiment was performed for four cycles within 10 sec. Experimental results for the grasping motion control are shown in Figure 6.20, where Finger 1 to Finger 5 are for thumb, index, middle, ring, and little fingers, respectively. The solid lines are the desired angles, which are the human finger angles measured by MIDS, for the control input. The lines with circle markers are the resultant joint angles of the robotic fingers, which are measured by the joint angle sensors inside the joints.

The process for this experiment is shown in Figure 6.21. First, both the human hand and the robotic hand were opened (Figure 6.21a). Then, the human fingers began to move toward a grasping motion, which was followed by the robotic hand (Figure 6.21b). Finally, the operator closed his hand and let the robotic hand followed this motion (Figure 6.21c) before he returned his fingers to their starting positions to perform the next grasping motion.

Comparing the results from MIDS and the sensors that were already on the robotic fingers, the desired angles for control input lead the resultant joint angles by about 0.3 sec, which is from the physical response times of the motors. Beside this time lag, the signal patterns of the finger joints are tracked to the operator's finger motion, which means that the robotic hand system can perform a good grasping motion by using MIDS to control them wirelessly.

In the second experiment, the operator tried to control the robotic hand to grasp a ball, which was located in a preset position. The experimental process is shown in Figure 6.22. The operator and the robotic hand were initially in start position (Figure 6.22a). Then the operator moved his hand, and the robotic hand followed this motion to move to the ball position (Figure 6.22b). The operator controlled the robotic hand to perform a grasping motion to grasp a ball and then held the ball moving back to the start position (Figure 6.22c and 6.22d). Then, the operator controlled the robotic hand back to the ball position and released the ball (Figure 6.22e). Finally, both the operator's hand and the robotic hand returned to the start position (Figure 6.22f). This test was performed to demonstrate that an operator could really control a robotic hand wireless to grasp and move an object using his own hand and arm motions.

The moving angle data from the second experiment are shown in Figure 6.23. The regions (a) to (f) in Figure 6.23, are the experimental results corresponding to the

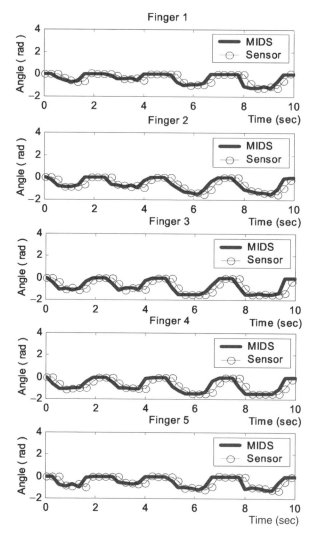

Figure 6.20. First experiment—angle data of five fingers from MIDS and joint angle sensors inside robotic hand system.

process states (a) to (f) in Figure 6.22. In the graph of Figure 6.23, the arm angle changed from 0rad to -3.14rad in region (a). This change means that the robotic hand changed from upward vertical orientation to downward orientation. In region (a), only the robotic arm moved, whereas all fingers did not move because the operator did not move his fingers. In region (b), the operator controlled the robotic finger to grasp a ball. The five fingers rotated from 0rad to -1.5rad. Similar to the first experiment, the MIDS command signals lead the robot finger signals by about 0.3 sec due to the physical response time of the joint motors. After that, the robotic

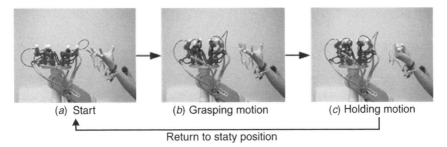

| (a) Start | (b) Grasping motion | (c) Holding motion |

Return to staty position

Figure 6.21. Process of grasping motion control of robotic hand system by MIDS.

hand held the ball and moved back to the starting position such that the arm angle changed from -3.14rad to 0rad in region (c). Then both the human hand and the robotic arm moved back to the ball position so the arm angle changed again from 0rad to -3.14rad [region (d)]. In region (e), the operator tried to control the robotic hand to release the ball; hence, the five fingers rotated from -1.5rad to 0rad. Finally, the arm angle changed back to the start orientation [from -3.14rad to 0rad (region (f))].

6.3.2 Computer Mouse on a Fingertip (MIDS-VM)

The mouse is one of the most pervasive devices in our lives. A recent survey of 1000 Internet users, fielded for Logitech by Greenfield Online, found that 63% of respondents spent more time holding their mouse than any other commonly held objects, including cell phones, remote controls, steering wheels, PDAs, or their lovers. The same survey found that 26% of respondents felt the mouse was the most important

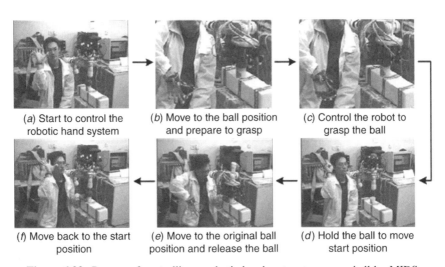

| (a) Start to control the robotic hand system | (b) Move to the ball position and prepare to grasp | (c) Control the robot to grasp the ball |
| (f) Move back to the start position | (e) Move to the original ball position and release the ball | (d) Hold the ball to move start position |

Figure 6.22. Process of controlling a robotic hand system to grasp a ball by MIDS.

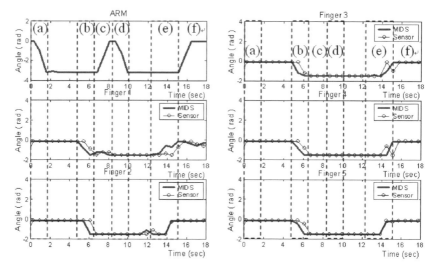

Figure 6.23. Second experiment—angle data from six MIDS sensors (five fingers and arm) and the joint angles sensors on the robotic fingers.

device in the computing experience, second only to the monitor (named by 51% of respondents).

Our proposed system, called Micro Input Devices System-Virtual Mouse (MIDS-VM), is made by merging MEMS sensors and a mechanical button system (Figure 6.24). It can be made into a ring-shaped structure or into any other structural forms. There are two designs for MIDS-VM: 1) button system and 2) 3D system.

For the button system, one two-axis acceleration sensor is placed inside the housing on the finger of the user in order to measure the motion or orientation of the fingertip and two buttons are placed on the housing to provide left and right click buttons. For the 3D system, only one 3D acceleration sensor is placed inside

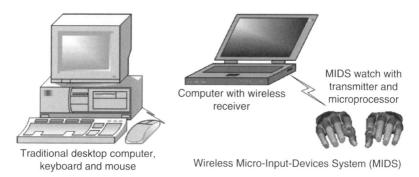

Figure 6.24. Configuration comparison between traditional input devices and a wireless MIDS.

the housing on the finger of the user in order to measure the 3D motion or orientation of the fingertip.

For both designs, a microcontroller is used to process and analyze sensor signals. It can be connected to a computer directly (for wired system), or it can be connected to a transmitter for wireless transmission. For the wireless system, the receiver is connected to the computer. The motion-to-command algorithms can convert the motions of the user to input commands so MIDS-VM can function as a PC mouse such that it allows the user to move curser, perform the drag-and-drop motion, left and right click, as well as control the scroll bar on the desk or in mid-air. Also, this novel computer–human interface allows a user to use a ring-shaped input device to handle graphical user interfaces. The details of the MIDS-VM are presented in the following subsections.

6.3.2.1 System Configuration. Based on the MIDS technology, the MIDS-VM can be developed as wired and wireless systems. For both wired and wireless systems, the MIDS-VM includes the sensing unit, signal processing unit, interface unit, and driver interface program. As mentioned, the MIDS-VM is a ring-shaped system. The MIDS units will be allocated based on the mechanical design of the ring. The detailed system configuration is discussed below.

6.3.2.2 MIDS Wired Ring. In the MIDS-VM wired ring, the sensing unit includes the motion sensor and buttons that are used to measure the human motion and button input. Two buttons in the MIDS-VM function as left and right click buttons. The sensing unit is installed in the ring-shaped housing, which was made by a rapid prototyping machine (FDM1600 by StrataSys Inc.). The MEMS sensor on a ring is placed on a PCB containing signal conditioning circuits. The signals from the sensor and buttons are directly transferred to the signal processing unit while the power is supplied by the signal processing unit through the wire. The schematic diagram for a MIDS-VM wired ring is shown in Figure 6.25.

The signal processing unit is integrated with the hardware interface unit (serial connection or USB connection). The two units are installed in the connector housing. For the serial system, the battery is needed to provide power to the system. The command is transferred to the computer through the serial port. For the USB system, no battery is needed. The system can get the power directly from the computer, and the signal command is transferred to it through the USB port. When the driver interface program detects that the commands are received, it will generate the command message to the operation system of the computer.

The MIDS-VM wired ring prototypes (for serial and USB system) are shown in Figure 6.26 and 6.27. Note that the design of the rings for the serial and USB systems is different.

6.3.2.3 MIDS Wireless Ring. In the MIDS-VM wireless ring, the sensing unit also includes the motion sensor and buttons. The two buttons function as left- and right-click buttons. The sensing unit is directly connected to the signal processing unit with the hardware interface unit (wireless transmitter module), and these units

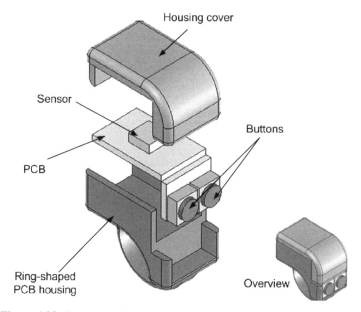

Figure 6.25. Schematic diagram of the MIDS-VM wired ring (two buttons).

are installed in the ring-shaped housing, which was made by a rapid prototyping machine (FDM1600 by StrataSys Inc.). The MEMS sensor on a ring, MCU, and RF transmitter are placed on a PCB containing signal conditioning circuits. The ring-shaped housing with the battery holder can carry the battery cell for supplying the power to the whole system. The schematic diagram for a MIDS-VM wired ring is shown in Figure 6.28. The entire MIDS-VM wireless ring prototypes are shown in Figures 6.29 and 6.30.

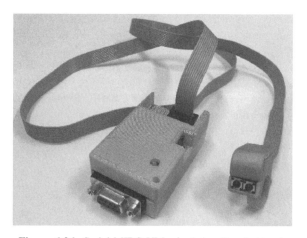

Figure 6.26. Serial MIDS-VM wired ring (two buttons).

Figure 6.27. USB MIDS-VM wired ring (two buttons).

The hardware interface unit including the wireless RF receiver can be connected to the computer through a serial or USB connection (see Figures 6.31 and 6.32). Similar to the wired systems, a battery is needed to provide power to the serial system, but no battery is needed for the USB system. The signals received from the RF receiver are transferred to the driver interface program, and then the driver program will generate the command message to the operating system of the computer.

6.3.2.4 Mechanical Design. Use of a computer mouse is not necessarily benign. Evidence is suggesting that the prolonged use of a computer mouse is associated with several upper extremity musculoskeletal disorders.

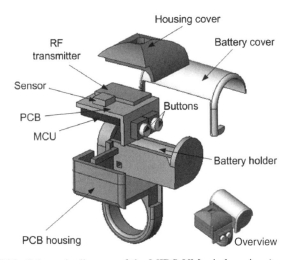

Figure 6.28. Schematic diagram of the MIDS-VM wireless ring (two buttons).

Figure 6.29. MIDS-VM wireless ring (side-button design).

As a human–machine interface, a good ergonomic design for MIDS is very important. There are two designs for the MIDS-VM mouse ring: side-button ring and top-button ring. The concept of side-button ring design is based on the direct pointing concept. People would like to point to the target with their finger. This side-button ring is convenient for directly pointing at the computer screen. However, it is not good in an ergonomic point of view. In the Figure 6.33, the force diagram of the side-button ring of MIDS-VM has been shown. To point to the target, the user needs to maintain the finger holding the ring horizontally (see Figure 6.33). Because of the weight of the ring and hand, the wrist has to pull the

Figure 6.30. MIDS-VM wireless ring (top-button design).

Figure 6.31. Prototype of serial receiver.

hand backward to balance the hand such that external force is needed. As a result, the user may feel tired when they use this side-button ring for a long time.

Because of the problems of the side-button ring, the top-button ring had been designed. It is also convenient for them to point to the computer screen, but it is not necessary for them to hold the finger horizontally. The user may keep the hand in a vertical position on the table; then the resultant force from the table support will directly balance out the weight of the ring and hand. No external force is needed to maintain the operation. As a result, the user can play with it for a much

Figure 6.32. Prototype of USB receiver.

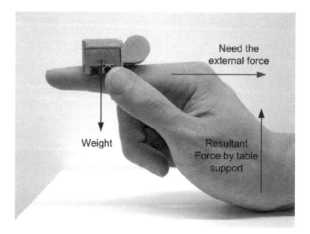

Figure 6.33. Force diagram of the side-button ring of MIDS-VM.

longer time than the side-button ring. The force diagram of the top-button ring is shown in Figure 6.34.

On the other hand, the ergonomic evaluation for button pressing has been conducted. The force diagrams for side-button ring and top-button ring are shown in Figures 6.35 and 6.36. It can be seen that the user tried to press the button on the side-button ring. A force was applied on the ring so that a force was needed against this force from the button pressing (see Figure 6.35). However, the user does not need to provide external force during the pressing of the button in the top-button ring because the resultant force from the table cancels out the force from button pressing. Therefore, the user can save energy by using the top-button ring, comparing it with the side-button ring.

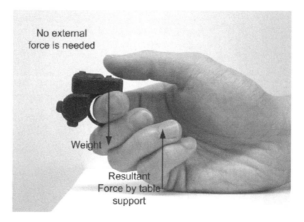

Figure 6.34. Force diagram of the top-button ring of MIDS-VM.

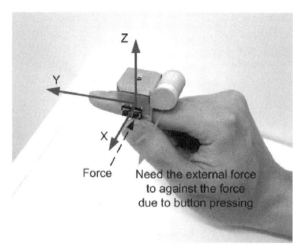

Figure 6.35. Force diagram for button pressing (side-button ring).

This ergonomic design of the top-button ring keeps the arm positioned for natural, comfortable movement, which has been clinically proven to be able to reduce muscle strain and discomfort associated with Carpal Tunnel Syndrome and repetitive stress injury, compared with the traditional mouse.

A pilot test has been conducted by the Sport and Physical Education Department of The Chinese University of Hong Kong. A female subject used a keyboard and MIDS-VM to play a computer game for around 1 hour. From the result of trend comparison, the muscle activities of the subject were larger when using the keyboard to perform the selected game task at flexor carpi ulnaris, extensor carpi ulnaris, and

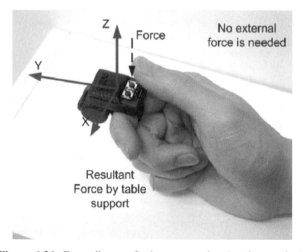

Figure 6.36. Force diagram for button pressing (top-button ring).

trapezius 1. Muscle fatigues might occur at flexor carpi ulnaris, trapezius 1, and trapezius 4 after 50 minutes. Although the sample size is too small, this pilot test can basically reflect that using MIDS-VM would be potentially better than the traditional input devices like the keyboard.

6.3.2.5 *Theory of Motion-to-Command Algorithm.* The working principle of MIDS-VM is based on the rotation of the user's finger/wrist. The motion-to-command algorithm of MIDS-VM is the conversion of tilt and roll angle to x and y movement on the screen (i.e., a mapping for rotation to linear movement). The consideration of this algorithm is to provide a smooth, easy-to-learn, and effective operation method to users. To achieve this goal, a nonlinear equation for rotation-to-linear motion has been developed.

$$D_n = a\left(\frac{D_{n-1}}{s_1}\right)^{P1} + b\left(\frac{D_{n-1}}{s_s}\right)^{P1} + cD_{n-1} + 1 \tag{6.19}$$

where D_n is the current movement (in pixel); D_{n-1} is the previous movement (in pixel); a, b, and c are the scaling factors. $P1$ and $P2$ are the power indices, and $S1$ and $S2$ are the scaling factors for the history data.

To make it easy to control, the algorithm should allow slight movement during the first period of time. In this study, we assumed the movement has not been too large in the first 0.6 sec such that the maximum movement should not exceed 200 pixels (which is about 25% of the width of a 800×600 screen size).

It is necessary to set an upper bound for the movement of the mouse cursor to avoid the "disappearance" of the cursor. Thus, the maximum movement has been set to prevent the cursor from moving too fast:

$$\text{If} \quad D_x > B_{\text{upper}}, D_x = B_{\text{upper}} \tag{6.20}$$

There are three constraints for setting the parameters:

1. At 0.5 sec, the mouse cursor movement should not exceed 200 pixels (in order to keep the motion slow and smooth for precise control).
2. From 0.5 sec to 1 sec, the mouse cursor movement should reach the maximum speed to provide a fast response.
3. At 1 sec, the mouse cursor should be moved around 800 pixels (assume 800 pixels is the maximum distance in a 800×600 screen size) such that the maximum speed or upper bound of the mouse movement is set to keep fast travel as well as to make the move visible.

By trial-and-error parameter setting, the parameters of the MIDS-VM motion-to-command algorithm are determined for fulfilling the constraints of the mouse movement. The values of the parameters are shown in Table 6.2.

Table 6.2. Parameters of MIDS-VM Motion-to-Command Algorithm

Parameter	Values	Parameters	Values
a	3.5	P1	6
b	1.2	P2	1.2
c	0.605	S1	2
B_{upper}	20	S2	3

Using the above parameter setting, the experiments were conducted to evaluate the movement of MIDS-VM. In Figure 6.37, the mouse movement operated by MIDS-VM has been shown. The mouse movement before 0.5 sec did not exceed 200 pixels, whereas the mouse can move up to around 800 pixels in 1 sec. The command for the mouse movement by MIDS is shown in Figure 6.38. The command reached the upper bound (20 pixels) after 0.5 sec in order to provide the maximum speed for the mouse movement. This experimental data show that the mouse movement is in an exponential curve to maintain the small movement at the first 0.5 sec (for precise control) and to provide large movement for fast mouse control after 0.5 sec.

6.3.2.6 *Working Principle.* To maintain the consistency of the MIDS products, the working principles for wired and wireless rings are the same. Based on the MIDS-tech mentioned, the data format is set to 3 bytes of data. The start byte is 255 to 252. The meanings of the start bye values can be found in Table 6.3. It has been set such that the final two bytes cannot be larger than 252. Thus, the acceleration value range

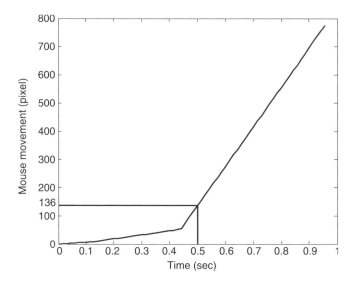

Figure 6.37. Mouse cursor movement operated by the MIDS-VM.

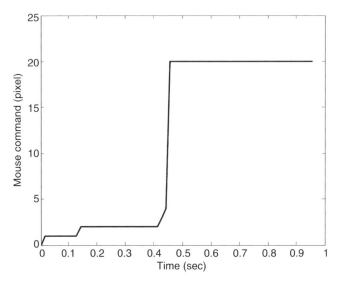

Figure 6.38. Command for the mouse movement by MIDS-VM.

in x- and y-direction is 0 to 251. Note that the acceleration is T1/T2, which is the ratio of the period of HIGH to the duty cycle. The actual value of the acceleration will be translated in the MIDS driver interface.

Since the MIDS-VM is based on the rotation of the user's finger/wrist (see Figure 6.39), the acceleration ranges from $-g$ to g such that the range of the output values, Yout and Xout, is 0 to 251. Zero gravity is represented by Aout = 126. The sequence of the output value, Aout, of the MIDS-VM is shown in Figure 6.40.

With the above output sequence, the driver interface receives the data through either the wired or the wireless receiver connected to the serial/USB port. Then the driver interface recognizes the start byte first (to check Aout > 251, if yes, then it is start byte) and classifies the button-pressing status based on Table 6.3. The following byte is stored in the variable Yout, and the next byte is stored in Xout. After receiving those data, they are passed to the motion-to-command algorithm to generate the commands. The details of the motion-to-command algorithm will be presented in the following section. The working principle is shown in Figure 6.41. When the MIDS mouse ring is rotated clockwise about the y-axis (the acceleration value Yout > 126), it is recognized as the "move cursor right" action.

Table 6.3. Meanings for Start Byte Values

Start Byte Values	Value Meaning
255	No button has been pressed
254	Left button has been pressed
253	Right button has been pressed
252	Two button have been pressed

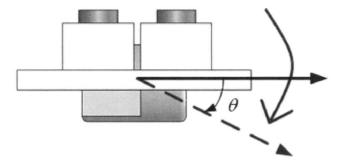

Figure 6.39. Working principle for MIDS-VM—Rotation of user's finger or wrist.

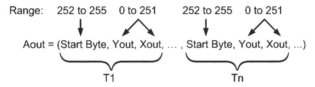

Figure 6.40. Output sequence of the MIDS-VM.

If it is rotated anti-clockwise (the acceleration value Yout < 126), it is recognized as the "move cursor left" action. Similarly, it is recognized as the "move cursor up" action when the MIDS mouse ring is rotated clockwise about the x-axis (the acceleration value Xout > 126). If rotated anti-clockwise (the acceleration value Xout < 126), it is recognized as the "move cursor down" action (see Figure 6.42 and Tables 6.4–6.6).

6.3.3 Computer Game Interaction Using MIDS

The realism and sophistication of computer games continue to increase as computer hardware becomes less expensive. New features for games nowadays are realistic 3D

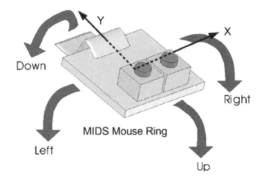

Figure 6.41. Working principle of the MIDS-VM.

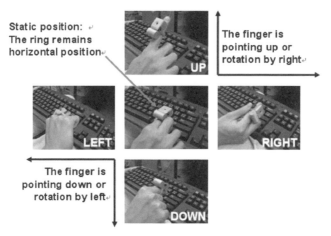

Figure 6.42. Demonstration for MIDS-VM.

scenes and live-like sounds. In addition, low-cost head-mounted displays will soon be available and will make truly immersive virtual reality experiences affordable in the near future. Both the gaming industry and users have become intrigued by 3D graphics technology, and a number of game software developers agree that the 3D graphics performance in mobile phones has finally surpassed that of the Game Boy Advance [14].

Mobile phones are not the only platform. In fact, a variety of other products are being announced. The N-Gage from Nokia Corp of Finland [15], and the Zodiac

Table 6.4. Technical Data of Wireless MIDS Mouse Ring

E-MIDS Acceleration Range ($1\,g = 9.81\,\mathrm{ms}^{-2}$)	$\pm 2\,g$
Acceleration resolution	$5\,\mathrm{mg}$
Angle range	$-0°$ to $360°$
Angle resolution	$1° \pm 0.4°$
Data sampling frequency	$70\,\mathrm{Hz}$
Command transmission rate	$2400\,\mathrm{bit/sec}$
Supply current	$\sim 8\,\mathrm{mA}$
Operating voltage	$3\,\mathrm{V}$
Power consumption	$\sim 24\,\mathrm{mW}$

Table 6.5. Technical Data of Wireless MIDS Receiver Interface

	Serial Port	USB Port
Supply current	$\sim 11.5\,\mathrm{mA}$	$\sim 31.5\,\mathrm{mA}$
Operating voltage	$5\,\mathrm{V}$	$5\,\mathrm{V}$
Power consumption	$\sim 57.5\,\mathrm{mW}$	$\sim 157.5\,\mathrm{mW}$

Table 6.6. Technical Data of Wired MIDS Mouse Ring

Acceleration Range ($1\,g = 9.81\,ms^{-2}$)	$\pm\,2\,g$
Acceleration resolution	$5\,mg$
Angle range	$-0°$ to $360°$
Angle resolution	$\$1 \pm 0.4°$
Data sampling frequency	$150\,Hz$
Command transmission rate	$9600\,bit/sec$
Supply current	$\sim30\,mA$
Operating voltage	$5\,V$
Power consumption	$\sim150\,mW\$$

from Tapwave Inc. of the United States all offer high-performance 3D graphics [16]. Around the end of 2004, Sony Computer Entertainment Inc. (SCE) of Japan released the PlayStation® Portable (PSP) [17], with a peak performance of 3.3 M-polygons/sec. While providing only half the 6.6 M-polygons/sec of the PlayStation®2, it still represented a stunning improvement over the first-generation PlayStation®.

To complement this trend, new input devices will be needed to improve the realism of the gaming experience. A standard joystick may no longer be an acceptable substitute for a gun, sword, fist, or steering wheel. Essential Reality is developing a new generation of product called P5 that enables the interface between human and machine in video gaming applications [18]. The P5 is a glove-like peripheral device based on proprietary bend sensor and remote tracking technologies that provides users with intuitive interaction with 3D and virtual games.

Using MIDS technology, a multifunctional controller can be developed. The same module can be applied to the control of PDAs and mobile phones gaming by the usage of a GUI. The idea of this multifunctional game controller is to make use of a motion sensing module with a microcontroller as a center control unit that is allowed to plug-in into different plastic housings to emulate different applications such as a joystick, gun, joy-pad, racing wheel, and others yet to be developed. This concept is illustrated in Figure 6.43.

6.3.3.1 System Configuration. The system requirement for gaming application is very high. Sensitivity, functions, and quality of the usage have to keep in high performance. As a new motion-based controller, more functions are needed. Based on the MIDS technology, the hardware for this game controller is similar to the MIDS-VM ring, but more buttons are added, and the driver and the GUI are modified specifically for the game application.

6.3.3.2 4-button MIDS Game Controller Ring (MIDS-GC ring). A four-button wired MIDS ring is designed with game control functions. The sensing unit includes the motion sensor and buttons that are used to measure the human motion and button input. Four buttons in the ring function as mouse buttons and/or keyboard buttons. The sensing unit is installed in the ring-shaped housing, which was made by a rapid prototyping machine (FDM1600 by StrataSys Inc.). The MEMS sensor on a

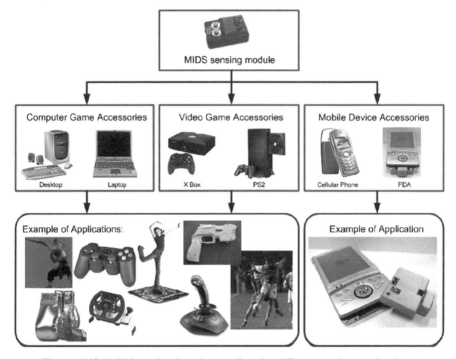

Figure 6.43. MIDS motion-based controllers for different gaming applications.

ring is placed on a PCB containing signal conditioning circuits. The signals from the sensor and buttons are directly transferred to the signal processing unit, whereas the power is supplied by the signal processing unit through the wire. The signal processing unit is integrated with the USB interface unit. The driver interface program has to be installed in the computer, which is used to receive the motion signal to and convert the command message to the operating system of the computer. The prototype of a four-button MIDS wired ring is shown in Figure 6.44.

6.3.3.3 Gaming Platforms. Two commercial computer game platforms are selected to demonstrate the functions of the MIDS-GC ring. Tom Clancy's Rainbow SixTM 3 - Raven Shield is a famous computer shooting game that required both a mouse and a keyboard to play. As a MIDS-GC ring can provide mouse and keyboard control, a user can just use one input device to play with this game.

Team17's Stunt GP is a nice 3D PC racing game that only required a keyboard to play. However, users always prefer to use the racing wheel because they will enjoy the game more (more like driving). A MIDS-GC ring can provide such a feature based on the similar operation principle of the driving wheel. A MIDS-GC can convert the motion and button input into the keyboard command so that user can drive the virtual racing car by rotating their hand. The function of the MIDS-GC ring is shown in Figure 6.45 and Table 6.7.

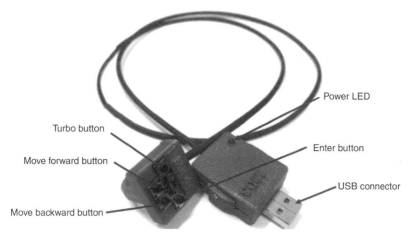

Figure 6.44. Function description of a four-button wired MIDS game controller ring.

6.3.3.4 Theory of Motion-to-Command Algorithm. The motion-to-command algorithm for MIDS-GC is same as MIDS-VM but with more functions because more buttons are added (Figure 6.46 and 6.47). The MIDS-GC user configuration interface allows the user to select the function for different motions. Mouse and keyboard functions are allowed to be selected.

6.3.3.5 Working Principle. The detail description for demonstrating the operation of an MIDS-GC ring for the game Stunt GP is shown below:

1. Turns left and right by hand motion (see Figure 6.48)
2. Other functions such as move forward, move backward, turbo accelerate, and enter are accessed by pressing the corresponding buttons on the MIDS racing ring.

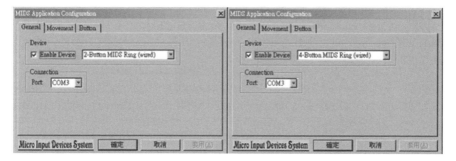

Figure 6.45. MIDS multifunction interface for two- and four-button ring.

Table 6.7. Technical Data of Wireless Game Controller Ring

E-MIDS Acceleration Range ($1\,g = 9.81\,\text{ms}^{-2}$)	$\pm 2g$
Acceleration resolution	$5\,\text{mg}$
Angle range	$-0°$ to $360°$
Angle resolution	$1° \pm 0.4°$
Data sampling frequency	$150\,\text{Hz}$
Command transmission rate	$2400\,\text{bit/sec}$
Supply current	$\sim 10\,\text{mA}$
Operating voltage	$3\,\text{V}$
Power consumption	$\sim 30\,\text{mW}$

6.3.4 MIDS for PDA Interaction (Embedded-MIDS: E-MIDS)

Besides the desktop and laptop computer, mobile computers are developing rapidly. Many people want to have a tiny device like a digital organizer to help them to manage the personal data. In addition, people start to think of playing mobile

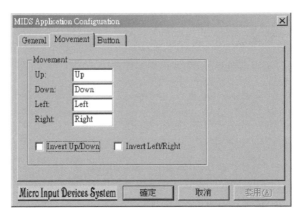

Figure 6.46. MIDS interface for motion settings.

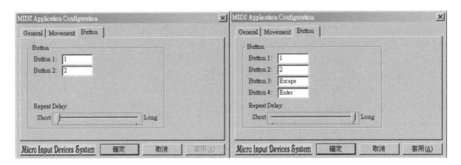

Figure 6.47. MIDS interface for button function selection.

Figure 6.48. To rotate the MIDS mouse ring in an anticlockwise direction.

games. With the advances of technologies in mobile phones and personal digital assistants (PDAs), more and more mobile devices can be used to play games. Thus, user-friendly input devices, which are also convenient to use on a mobile platform, are very important for playing games on PDAs or mobile phones. As a game controller, the system requirement (high sensitivity and user-friendly control, etc.) is much higher than other input devices. A novel mobile input device or game controller will potentially drive the game development for mobile devices. Various input devices (such as joystick, joy pad, and racing wheel) other than a traditional mouse and keyboard have been already developed for desktop computers.

However, the most common input devices for PDAs are still stylus and keypad. They have survived without significant changes even though the PDAs have already evolved several generations. Stylus is small in size and easy to carry. Nevertheless, it may destroy the touch screen easily, especially by users who are overly excited while playing games. To protect the screen, a stylus-like product, Magic Finger [19], is available on the market. Magic finger is a rubber pack that is worn on the finger and allows users to point at the screen. Although it can prevent the damage of the screen, it is obviously not suitable for game playing. For some games such as role-play games (RPG), stylus and magic finger are also not suitable to navigate the character around the environment because the users' fingers will block the screen, which will cause an inconvenience while controlling the game. Moreover, such input devices are too simple and have limited functions. Thus, game developers are currently constrained by the stylus in creating games for mobile devices.

Many mobile games use keys as controllers. However, because of the compact size of the PDAs, the build-in keys are always very small and difficult to access. To tackle this problem, manufacturers have tried to minimize the external keyboard size and develop many compact products such as wireless keyboards [20] or folding keyboards [21]. However, it is still a burden for the users to carry it around. (Normally, the external keyboards are the same size as a PDA.) Furthermore, it is inconvenient for users to play games with external keyboards on a bus or walking on a street.

Besides the above disadvantages of existing input devices for mobile products, the users usually cannot fully enjoy the games. Many games such as action games and sports games could be much more exciting if the players can use their body motion. This is the reason why many people are crazy about playing virtual reality games.

On the other hand, the functionality of mobile phones and PDAs is ever increasing while their size is ever decreasing. These trends will limit the usability of the devices if traditional user interfaces such as buttons and jog dials are not replaced by the next generation of "natural" user interfaces such as speech and motion. Considering the needs of the mobile devices, a novel controller called E-MIDS has been invented to take advantage of the human mobility as well as motion capturing for game control. In addition, it is a small-size, easy-to-carry, user-friendly, and multifunctional controller that is suitable for mobile devices (the external device is just one third the size of a PDA). Moreover, E-MIDS can be either plugged-in or embedded into mobile devices such as mobile phones and PDAs, or by replacing their existing housing structure. The conceptual system configuration of E-MIDS for mobile devices is shown in Figure 6.49.

With the embedded feature, users just need to take care of the PDAs and there is no need to carry any other external devices. As the users only have to move or rotate their PDAs to control the game, it will not damage any parts (such as the screen) of the PDAs. Most importantly, it provides a way for the users to control the games with their body motion, and lets the users enjoy the games more. The comparison between E-MIDS and existing input devices for PDAs is shown in Table 6.8.

Based on MIDS technology, E-MIDS is invented that can be externally connected to or internally embedded in mobile devices such as PDAs or cellular phones. For external E-MIDS, it can be connected to mobile devices through the serial port or USB port. For internal E-MIDS, it is embedded inside the mobile devices, which means that E-MIDS will not be shown to users. Using E-MIDS, the users can use their motion as an input method to play or control the mobile games. For these two systems, E-MIDS can get the power supply from the mobile devices directly. In the coming sections, the demonstration for an external E-MIDS connected to the PDA, Sharp Zaurus, will be illustrated.

6.3.4.1 System Configuration. For the demonstration of using E-MIDS to control a mobile device, the targeted mobile device is the PDA Zaurus SL5500 manufactured by Sharp. The game control realized for this demonstration system is "Aliens."

The system configuration and the prototype of E-MIDS are shown in Figure 6.50. There is a dual-axis MEMS sensor (the sensing unit) that is used to measure the human motion. In addition, there are two buttons on the PCB. The users can play games not only with their motions, but also with the traditional method of pressing buttons.

Since E-MIDS is connected to the PDA externally as is shown in Figure 6.51, the sensor can measure the motion of the PDA when the user is holding it. The MCU (the signal processing unit) will get the input data both from the sensor and the buttons and then it will process the signal. Through the motion-to-command algorithm, the motion signals will be converted to command signals. After that, the commands will be converted to serial format by the RS232 chip. Finally, the serial format

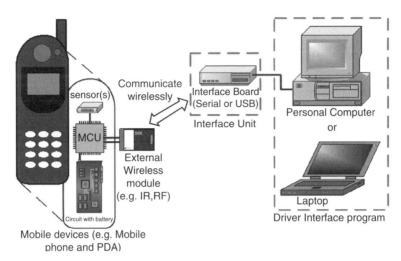

Figure 6.49. The conceptual system configuration of E-MIDS for mobile devices.

commands will be sent to the PDA to control the game. The data flow of E-MIDS is shown in Figure 6.52.

In this work, the Sharp Zaurus SL5500 is selected as the host device in the demonstration. This system provides a serial/USB IO port at the bottom part of it as shown in Figure 6.51. For the hardware configuration, the serial connecter of the E-MIDS is connected to the Zaurus IO port. For the software driver, the Zaurus (the operating system is Linux) is booted in "INIT 3" mode to activate the serial function of the Zaurus's IO port, which can be accessed via "/dev/ttyS0."

6.3.4.2 *Theory of Motion-to-Command Algorithm.* The program flow of most games for Zaurus is shown in Figure 6.53. A top-level class is instantiated at the beginning of the game and handles the game and display control. To allow E-MIDS to directly control games hosted on the Zaurus, a mechanism is added to read the E-MIDS input signals, and then to pass it to the data handling routine. The data handling routine can further interpret the signals according to the game semantics, and then it generates the corresponding game commands. This mechanism

Table 6.8. Comparison Between E-MIDS and Existing Input Devices for PDAs

	Size	Easy to Carry	Function	Damage PDA
Stylus	Small	Yes	Simple	Yes
Keypad	Medium	Yes	Multi	No
Magic finger	Small	Yes	Simple	No
External keyboard	Large	No	Multi	No
External E-MIDS	Medium	Yes	Multi	No
E-MIDS	Nil	Nil	Multi	No

Figure 6.50. Schematic and prototype of E-MIDS.

is added to the top-level class. Thus E-MIDS can control the game via this top-level class and game designers need not modify the implementation of the game. This motion-to-command algorithm in the game "Aliens" is shown in Figure 6.54.

In Figure 6.55, the black dotted line is the motion-to-command algorithm of E-MIDS added to the original "Aliens" game program flow. The data handling is event driven. At the start of the game, the serial port "dev/ttyS0" is opened and configured to receive the E-MIDS signals. Then a socket notifier [22] is created and bounded to it. The socket notifier will notify the data handling routine when the data from E-MIDS are received. In the data handling routine, the validity of the data is checked first. The data are further interpreted only when it is correct. The E-MIDS signals consist of motion information as well as the E-MIDS button status. They are handled separately.

6.3.4.3 Working Principle. The motion information, which is measured by the E-MIDS sensor, is converted to the pitch angle θ. The pitch angle (the range is from

Figure 6.51. Installation of E-MIDS for a Sharp PDA.

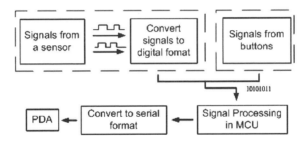

Figure 6.52. Data flowchart of E-MIDS.

$-90°$ to $90°$) of the E-MIDS is relative to the world coordinate frame as shown in Figure 6.55. Based on the angle θ, the rotational motion of the user's hand can be inferred so that the command for the game control can be defined as listed in Table 6.9. According to the results of trial-and-error testing, the threshold region is defined between $-7°$ to $7°$. When the θ is within this region, the PDA is regarded as stationary in order to ignore the users' hand shaking. In the original "Aliens" game, the "LEFT" and "RIGHT" direction keys are used to move the spaceship leftward and rightward, respectively. Therefore, when the pitch angle θ is between $-90°$ and $-7°$ (the user rotated the PDA counterclockwise by θ), a left key pressed event is sent to the top-level class to move the spaceship to left. Similarly, when the pitch angle θ is between $7°$ and $90°$ (the user rotated the PDA clockwise by θ), a right key pressed event is sent to move the spaceship to the right.

On the other hand, the spacebar is used to fire. Thus a space key pressed event is sent to activate the fire function when a button click action (E-MIDS buttons) is received.

Note that the E-MIDS motion-to-command algorithm (see Figure 6.54, the black dotted line) is a module added to the original program flow. It does not affect the performance of the original game design. This module feature can provide an easy way

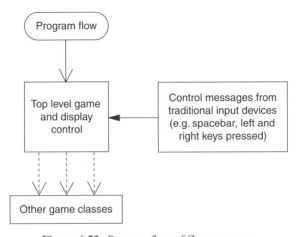

Figure 6.53. Program flow of Zaurus games.

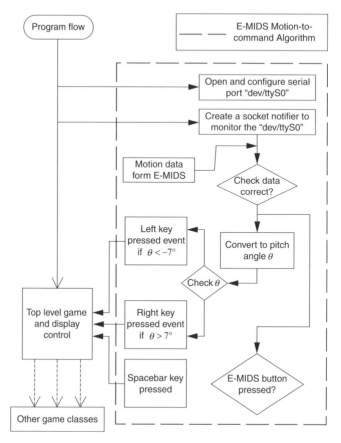

Figure 6.54. Program flow of "Aliens" with E-MIDS motion-to-command algorithm.

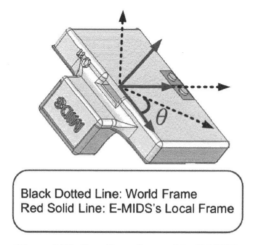

Figure 6.55. Coordinate frame of the E-MIDS.

Table 6.9. Illustration of Motion-to-Command Algorithm

Angle (α)	Rotate Motion	Command
$-90° < \theta < -7°$	Counterclockwise	Left direction
$-7° \leq \theta \leq 7°$	Stationary	No change
$7° < \theta < 90°$	Clockwise	Right direction

for the game designer to use the E-MIDS as a game controller because this motion-to-command algorithm can be separately developed by just knowing how to play the game and there is no need to study the details of the game implementation.

The demonstration of playing the "Aliens" with E-MIDS is shown in Figure 6.56. When the user rotates the PDA in a clockwise direction, the spaceship will move leftward as in Figure 6.57. Similarly, when the user rotates the PDA in counterclockwise direction, the spaceship will move rightward. There is no denying that "Aliens" will become more entertaining compared with key-controlled "Aliens" (which keeps the players busy pressing buttons on the small keypad on the PDA) and stylus-controlled "Aliens" (which not only keeps the players tracking the spaceship but also causes much confusion and many mistakes). Therefore, E-MIDS can make the control of the "Aliens" become more direct by using the hand motion.

The performances of E-MIDS and traditional input devices in playing "Aliens" are evaluated and compared. The transmission rate of E-MIDS data to Zaurus is around 70 commands per second. It means that the sensitivity is high enough to measure the human hand motion (normally the rate of human hand motion is less than 10 Hz without considering the hand shaking). Therefore, E-MIDS can respond fast enough to human motion. As a result, the factor that affects the performance of E-MIDS and traditional input devices in playing a game is the user's habit, which is how fast the user can input commands with those devices.

An experiment is designed and carried out to measure the user's habit of using E-MIDS and traditional input devices in playing "Aliens." There are three commands to "Aliens": move left, move right, and fire. For the fire command, E-MIDS is based on button pressing, whereas traditional input devices are just based on key pressing such that they make no differences. Thus, only the move left and right commands need to be investigated. To get how fast the user inputs commands, the time when the user inputs each command has to be recorded. As shown in Figure 6.58, an additional block is inserted into the game flow to record the time of a user's input command.

An experienced user was invited to play "Aliens" with both E-MIDS and traditional input devices. He was asked to generate move left and right commands alternatively as fast as he can by using E-MIDS, keypad, and stylus, respectively. For E-MIDS, the user rotated the PDA by his wrist to generate game commands. For keypad, the user pressed the left and right keys with thumb. For stylus, the user pointed to the left side of the spaceship to move it left and pointed to the right side to move it right. In each experiment, the time for input generation and the number of input cycle are recorded for 1 minute (60 sec).

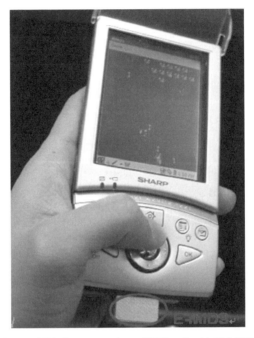

Figure 6.56. Demonstration of "Aliens" with E-MIDS.

The total numbers of input cycles in 1 minute for three input devices are listed in Table 6.10. The user using E-MIDS can generate game commands two times faster than using a keypad, and almost three times faster than using a stylus. As described before, the threshold region for the pitch angle q is defined between $-7°$ and $7°$ to ignore a user's hand shaking, and this threshold has been experimentally proved to

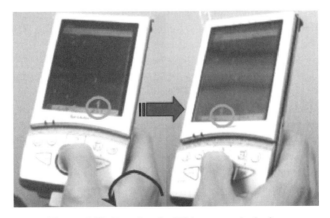

Figure 6.57. Rotating the PDA counterclockwise.

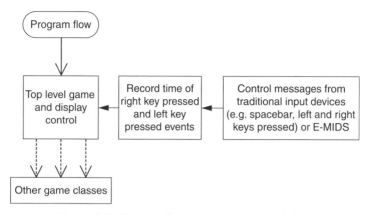

Figure 6.58. Program flow to capture command time.

be able to eliminate most hand shaking. However, the setting of the threshold region reduced the sensitivity of the control. Fortunately, the results of this experiment show that the response of control is faster than that of the traditional input devices. It means that E-MIDS can provide faster response time and eliminate the error from hand shaking. Based on the experimental results of the above experiment, the frequency response of each input method is further investigated.

The frequency response measurement is based on the power spectrum density (PSB) of the input rates by using E-MIDS, keypad, and stylus (see Figure 6.59). The PSD of the input rates of different input devices shows the dominating frequency (peak rate) for E-MIDS is 3.52 Hz, whereas the dominating frequencies for keypad and stylus are 1.95 Hz and 1.56 Hz, respectively. The components of E-MIDS induced by hand shaking are always of high frequency if there are any. Therefore the peak rate 3.52 Hz can more accurately reflect the user's intentional hand motion. This frequency response further proved that with E-MIDS the user can generate a higher frequency of input commands.

According to the experimental results, E-MIDS can provide a fast response for generating input commands that is two times faster than the traditional input devices. From the demonstration, E-MIDS is proved to be stable, controllable, easy-to-use, and user-friendly, which totally fulfilled the requirements of mobile-based gaming and is more suitable than the existing input devices for the mobile products. With E-MIDS, a new generation of mobile games will come in the near future because E-MIDS can provide more functions and higher flexibility for game designers (Table 6.11).

Table 6.10. Total Number of Input Cycles in 1 minute

	E-MIDS	Keypad	Stylus
Number of input cycles	267.5	118.5	93.5

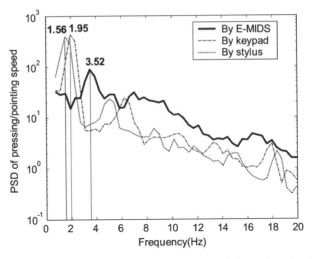

Figure 6.59. PSD of input rates by using E-MIDS, keypad, and stylus.

6.4   CONCLUSION

In this chapter, a novel MIDS technology is invented and it is a mature technology that makes use of human motion as a control methodology to develop different novel systems for computer input devices, motion sensing devices, and motion-based controllers. A series of workable system designs and algorithm/software developments for the human–robot–hand interaction system, fingertip computer mouse, ring-shaped game controller, and PDA controller are achieved.

With MIDS technology, a new generation of computer control style will come in the near future. You can imagine that the monitor can combine with the computer processor such that it is wearable (or just like glasses). This computer with MIDS can let users work everywhere when they are available. That means the office can be anywhere and that any time can be a working hour. It may bring us the revolution in computer development and human–machine interface. Therefore, it is concluded that this novel MIDS technology has the potential to give a deep positive impact to us in daily life. Besides the applications of MIDS mentioned above (such as computer input device, motion sensing device, and robot control device), there are many potential applications for this invention. Piano players can use the MIDS to play a virtual piano. Their piano keyboard input patterns can be transferred to the computer, which could be used for song recording or live-performance. Another application is that users can use the MIDS to interface with a PC using conventional sign-language gestures; i.e., users can simply use sign-language motions, and the computer will be able to translate the motions into sentences without any keyboard typing. One of the most meaningful applications for the MIDS will be its usefulness to help the blind. MIDS can help blind people to make brail-based typing by using different finger motion patterns. When the finger motion patterns are transferred to the computer, the computer

Table 6.11. Technical Data of the E-MIDS

E-MIDS Acceleration Range ($1\,g = 9.81\,ms^{-2}$)	$\pm 2\,g$
Acceleration resolution	$5\,mg$
Angle range	$-90°$ to $90°$
Angle resolution	$1° \pm 0.4°$
Data sampling frequency	$70\,Hz$
Command transmission rate	$2400\,bit/sec$
Supply current	$\sim 7\,mA$
Operating voltage	$3\,V$
Power consumption	$\sim 20\,mW$

can translate them to words and then save the words as a document. Other applications for the MIDS will include emulation of a laser-pointer, and as a light pen for languages such as Chinese.

Actually, MIDS technology can be applied on all areas related to motion. Anyone may think of hundreds of applications using MIDS. The objective of this chapter is to introduce the fundamental knowledge of MIDS technology and provide an alternative for the people who want to make use of the human motion as a control method or to build an effective motion sensing system for motion measurement.

REFERENCES

1. Maxim Integrated Products Inc..: http://www.maxim-ic.com/
2. Future Technology Devices International Limited: http://www.ftdichip.com.
3. Analog device: http://www.analog.com.
4. Memsic: http://www.memsic.com/.
5. STMicroelectronics: http://www.st.com/.
6. S. M. Sze, *Semiconductor Sensors*, Hoboken, NJ, Wiley-Interscience, 1994.
7. Atmel Corporation: http://www.atmel.com/.
8. Linx Technologies: http://www.linxtechnologies.com/.
9. ASIMO: http://world.honda.com/ASIMO/.
10. Sony QRIO Robot: http://www.sony.net/SonyInfo/QRIO/top_nf.html.
11. D. Ding, Y.-H. Liu, J. Zhang, and A. Knoll, "Computation of fingertip positions for a form-closure grasp," in *Proc. of 2001 IEEE International Conference on Robotics and Automation*, May 2001, Vol. 3, pp. 2217–2222.
12. Y. Hasegawa, J. Matsuno, and T. Fukuda, "Regrasping behavior generation for rectangular solid object," in *Proc. of 2000 IEEE International Conference on Robotics and Automation*, April 2000, Vol. 4, pp. 3567–3572.
13. C.C. Cheah, H.Y. Han, S. Kawamura, and S. Arimoto, "Grasping and position control for multi-fingered robot hands with uncertain Jacobian matrices," in *Proc. of 1998 IEEE International Conference on Robotics and Automation*, May 1998, Vol. 3, pp. 2403–2408.

14. Gameboy Advance: http://www.gameboy.com/.

15. Nokia's N-Gage: http://www.n-gage.com/.

16. Tapwave's Zodiac: http://www.tapwave.com/.

17. Sony's Playstation Portable: http://www.theregister.co.uk/2003/11/05/sony_playstation_portable_pics_pop/.

18. Essential Reality: http://www.essentialreality.com.

19. Kirrio Magic Finger: http://www.bargainpda.com/price/default.asp?productID=380&brandName= Kirrio&productName=Magic+Finger&display=productDetail.

20. PalmOne Infrared Wireless Keyboard for Palm PDAs – Palm Electronics: http://www.boondocksnet.com/cb/apfh-item_id-B0000E0ZY9-search_type-AsinSearch-locale-us.html.

21. Viewsonic Pocket PC V35 Foldable Keyboar, PPC-KYB-001, PPCKYB001: http://www.wholesalers-direct.com/ppckyb001.html.

22. Qt Toolkit – QSocketNotifier Class: http://doc.trolltech.com/qtopia1.6/html/qsocketnotifier.html.

CHAPTER 7

Ubiquitous 3D Digital Writing Instrument

7.1 INTRODUCTION

A MAG-μIMU, which is based on Micro ElectroMechanical Systems (MEMS) gyroscopes, accelerometers, and magnetometers, is developed for real-time estimation of human hand motions. Appropriate filtering, transformation, and sensor fusion techniques are combined in the Ubiquitous Digital Writing Instrument to record handwriting on any surface. An extended Kalman filter (EKF) based on a MAG-μIMU (micro inertial measurement unit with magnetometers) is designed for real-time attitude tracking and is implemented to record handwriting.

In classic IMU applications, the Kalman filter uses the gyroscope propagation for transient updates and correction by reference field sensors, such as gravity sensors, magnetometers, or star trackers. A process model is derived to separate sensor bias and to minimize wide-band noise. The attitude calculation is based on quaternion, which when compared with Euler angles, has no singularity problem. According to this filter framework, a Complementary Attitude EKF is designed by integrating the measurement updates from accelerometers and magnetometers alternatively, in order to compensate the attitude observation error caused by sensor limitations, such as inertial accelerations, magnetic field distortion, and attitude ambiguity along each reference field. Testing with synthetic data and actual sensor data proved the filter will rapidly converge and accurately track the rigid-body attitude. The pen-tip trajectory in space can be calculated in real time based on the real-time attitude estimation. Our goal is to implement this algorithm for motion recognition of a three-dimensional (3D) ubiquitous digital pen.

Three-dimensional position tracking can be directly obtained successfully by some existing technologies. For instance, ultrasonic and infrared positioning systems with high accuracy have been developed for tracking handwriting motions, such as the Mimio [1] and e-Beam systems [2]. The basic operating principle of these systems is that, when two transceivers are fixed on white board as the position reference and broadcast ultrasonic or infrared timing signals, the two transceivers can

Intelligent Wearable Interfaces, by Yangsheng Xu, Wen J. Li, and Ka Keung C. Lee
Copyright © 2008 John Wiley & Sons, Inc.

measure the two distances to the reflector, respectively, by detecting the phase difference after receiving the echo wave from the reflector on the pen-tip,. This source positioning technology is high in accuracy and quick in response. However, the system cannot track multiple objects simultaneously, and it requires a set of receivers and a transceiver to operate, which is not convenient for users.

In our ubiquitous 3D digital writing system, the overall goal is to develop a system to track handwriting motions without the need for wave sources such as ultrasonic or infrared signals. In sourceless navigation technology, such as inertial kinematics theory, an accurate attitude is fundamental for determining and keeping track of the rigid-body position in space. Because of nonlinearity in the system dynamic equations, bias error and random walk noise from attitude sensors will be accumulated and magnified, which leads to nonlinear distortions in position tracking.

Attitude tracking is widely used in navigation, robotics, and virtual reality. Classically, the problem of distortions in position tracking is addressed by the Attitude Heading Reference System (AHRS) [3, 4]. The AHRS uses gyroscope propagation for transient updates and correction by reference field sensors. However, classically, the performance is ensured by extremely accurate sensors and hardware filters. Because of its expensive cost and large system size, the AHRS has been limited in applications, especially for mobile human position tracking applications.

With MEMS sensing technology, the inertial sensors could be built with low cost and small sizes. However, they suffer in accuracy when compared with bulkier sensors. Nevertheless, new reliable attitude tracking systems have been developed based on low-cost gyroscope sensors and on the Global Positioning System (GPS). For feedback correction, Euler angles are derived from GPS to represent spatial rotation and a Kalman filter is implemented to fuse with the attitude propagation. But GPS signals are not available for indoor applications and the GPS attitude has a resolution limit for the handwriting application [5].

For static applications such as the unmanned ground vehicle (UGV) control [6], the MEMS accelerometers work reliable as gravity sensors. Euler angles can be derived directly [7]. However, during a dynamic situation, the accelerometer measurements for the gravitational accelerations will be interfered with the inertial accelerations, which then cannot be trusted for attitude. Furthermore, the pitch attitude along the gravity axis cannot be determined.

Magnetometers experience no such crosstalk disturbance in both situations. However, following the same approach, attitude ambiguity occurs along the magnetic field direction and Euler angles cannot be derived directly. Furthermore, the Earth magnetic field is overlapped by random noise from electromagnetic interference (EMI).

A Ubiquitous Digital Writing Instrument has been developed by our group to capture and record human handwriting or drawing motions in real time based on a MEMS μIMU [8]. However, position tracking using this μIMU· is not accurate because of sensor measurement noise and drifts.

Thus, an extended Kalman filter is designed to improve the system measurement accuracy of an integrated gyroscope and magnetometer device (MAG-μIMU) [9]. For the digital writing instrument application, the MAG-μIMU is affixed on a commercially available marker. During input hand motions, the filter tracks the real-time attitude of the pen with sensor bias separation and sensor random noise minimization. The attitude calculation is totally based on quaternion, which computes faster and has no singularity problem compared with Euler angles. This filter also applies to other reference field sensors for feedback, such as accelerometers or star trackers.

Based on this filter structure, a complementary extended Kalman filter for attitude is proposed to combine the accelerometers and magnetometers in the measurement update process. During the pause phase in handwriting, the complementary filter uses the accelerometers as gravity tilter for attitude reference feedback. During the writing phase, the measurement model switches to the magnetometers to avoid attitude error caused by the inertial accelerations in gravity sensor, by assuming the magnetic field distortion is tolerable within one stroke. This measurement mixture can compensate attitude ambiguity along the reference field direction for each other and rectify EMI in Earth's magnetic field.

7.2 HARDWARE DESIGN

Figure 7.1 illustrates the block diagram of a wireless MAG-μIMU with the real-time attitude filter system. The system can be divided into two parts. The first part is the hardware for the wireless sensing unit. The other part is the software for data access and 3D rotation sensing algorithms.

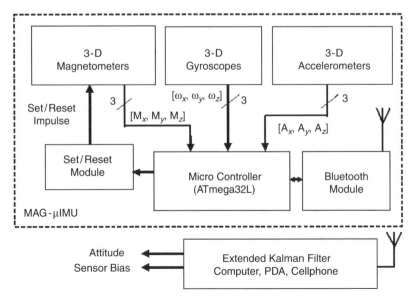

Figure 7.1. Wireless MAG-μMU block diagram.

The MAG-μIMU is developed for a wireless digital writing instrument and is used to record human handwriting. The MAG-μIMU is a hybrid sensing system with inertial sensors and magnetometers. The "μIMU" integrates the 3D accelerometers [10] and 3D gyroscopes [11] with strapdown installation [8]. The 3D magnetometers, "MAG" sensors, are added to measure the Earth's magnetic field [12–14]. The sensor unit is affixed on a commercially available marker to measure the inertial and magnetic information in the pen's body frame.

The output signals of the accelerometers $[A_x, A_y, A_z]$ and the gyroscopes $[\omega_x, \omega_y, \omega_z]$, which are the body frame accelerations and the roll, pitch, and yaw angular rates, respectively, are measured directly with an Atmega32L A/D converter microcontroller [15, 16]. The serial Bluetooth transceiver is implemented via a USART connection with the MCU for wireless communications [17, 18]. The serial USB transceiver is integrated for transfer backup and hardware development [19]. The digital sample rate of the sensor unit is 200 Hz, and the transmit baud rate is 57.6 kbps via Bluetooth wireless connection, which ensures rapid response to human handwriting.

Figure 7.2 shows the MAG-μIMU with strapdown gyroscopes and magnetometers for attitude tracking test. The sensor system uses four-layer printed circuit board techniques for noise reduction. The dimensions are within $56 \times 23 \times 15$ mm.

7.3 SIGNAL PROCESSING AND ANALYSIS

The stroke segment algorithm is proposed by Ref. [7] to separate the continuous pen-tip movement of character writing into independent strokes. The velocity of the pen-tip equals zero at the turning point between two consecutive strokes. This motionless phase in handwriting is important for compensating the error in attitude estimation and the position integration.

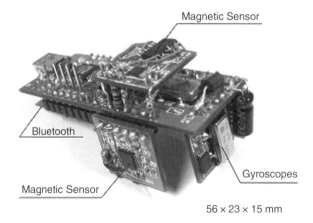

Magnetic Sensor

Bluetooth

Magnetic Sensor

Gyroscopes

$56 \times 23 \times 15$ mm

Figure 7.2. The prototype of the MAG-μIMU with a Bluetooth module.

In Ref. [7], the stroke segment is implemented based on the sample variance of inertial measurements. In probability theory, the variance of a random variable is a measure of the spread around its expected value. The sample variance is the approximation for variance with finite size N in real application. For the N sample set, the sample variance equals the unbiased mean square error, which indicates the scale of values of the sample set. Hence, the intensity of handwriting motion can be described by the sample variance of the corresponding acceleration sample set in real time.

For computational efficiency, a first-in first-out (FIFO) stack is preferred to store and process the successive sample data. The sample variance with window length N can be calculated as

$$\sigma = \sqrt{\frac{1}{N}\sum_{i=1}^{N}(x_i - \bar{x})^2} \tag{7.1}$$

where $\bar{x}$ is the sample mean

$$\bar{x} = \frac{1}{N}\sum_{i=1}^{N}x_i \tag{7.2}$$

For a sampled time instance k, if $\sigma^S_{|A_n|}(k) < \sigma_{th}$ for $k = k, k+1, \cdots, k+H$, then k is the beginning k_1 of the stroke.

For a sampled time instance $k > k_1 + W$, if $\sigma^S_{|A_n|}(k) > \sigma_{th}$ for $k = k, k+1, \ldots, k+H$, then k is the end k_2 of this stroke.

Where $\sigma^S_{|A_n|}(k)$ is a standard variance of $|A_n|$ over S samples until k, σ_{th} is a threshold of the standard deviation. W is the minimum number of samples for writing a stroke, and H is the minimum number of samples to keep $\sigma^S_{|A_n|}(k)$ from being less than σ_{th}.

As the unbiased motion intensity estimator of sample variance, the stroke segment algorithm does not have the concern of gravitational accelerations, which is changing with pen attitude. The sample variance evaluates in stochastics from the entire sample set block. This macroscopic information can efficiently discriminate hand motions by writing or trembling. However, the σ_{th}, W, and H should be predefined. These parameters have to try and test in calibration. The recognition efficiency may vary with characters of different size, especially for a different experimenter when the handwriting habit is different.

As the approximation to the expected variance, the error in sample variance caused in sampling may magnify the motion segment error when the sample set data are symmetric near the inflexion point as illustrated in Figure 7.3.

Suppose the acceleration signal for handwriting is a one-interval sine wave of 2.5 Hz as demonstrated on the blue curve in Figure 7.3. And the sample rate of the digital system is 50 Hz, which satisfies the Nyquist sampling theorem. The window

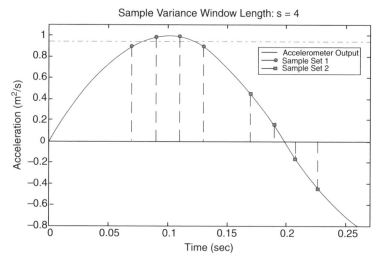

Figure 7.3. Acceleration variance underestimates motion intensity (velocity) caused by sampling error.

length for sample variance is 4 for a quick response. The sample variance for the first sample set equals 0.0485. On the other hand, after one measurement, the sample variance for the second sample set equals 0.3394. A trivial hand tremble signal simulated on the black line has the same sample variance as the first sample set. However, the pen-tip is accelerating during the first sample set around 0.1 seconds until the stroke ended at 0.4 seconds when the velocity decelerates to zero.

Thus, a Kalman filter is proposed to make use of the posterior motion information to address this problem. In a time update, the variance of acceleration is propagated as the system state as follows:

$$\hat{x}_k^- = \hat{x}_{k-1} \tag{7.3}$$

$$P_k^- = P_{k-1} + Q \tag{7.4}$$

In a measurement update, the estimation is updated by the sample variance $O(y_{k-N+1}, \ldots, y_k)$ as follows:

$$K_k = \frac{P_k^-}{P_k^- + R} \tag{7.5}$$

$$\hat{x}_k = \hat{x}_k^- + K_k(O(y_{k-N+1}, \ldots, y_k) - \hat{x}_k^-) \tag{7.6}$$

$$P_k = (I - K_k)P_k^- = \frac{R}{P_k^- + R}P_k^- \tag{7.7}$$

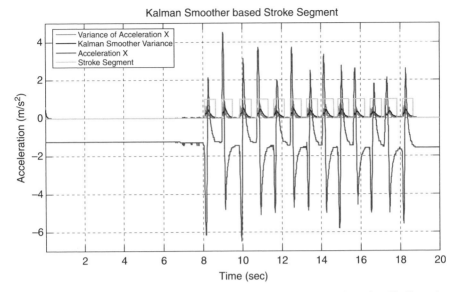

Figure 7.4. Kalman smoother-based stroke segment for motion status detection (1). See color insert.

where $O(y_{k-N+1}, \ldots, y_k)$ is the sample variance for the N data set until current k.

$$O(y_{k-N+1}, \ldots, y_k) = \sqrt{\frac{1}{N} \sum_{i=k-N+1}^{k} \left(y_i - \frac{1}{N} \sum_{j=k-N+1}^{k} y_j \right)^2} \qquad (7.8)$$

Figure 7.4 demonstrates the experiment result to verify the segment performance. A zigzag line with 13 strokes is written to verify the response and stability of the algorithm. The window length is set to: $N = 5$.

The stroke segment detail is shown in Figure 7.5 below. The Kalman filter tracks the sample variance with a quick reaction corresponding to the start and end of each stroke. The variance drop at a symmetric sample set near the maximum value can be compensated by previous estimation.

7.3.1 Kalman Filtering for MEMS Sensors

The extended Kalman filter consists of two stages. In the time update stage, the quaternion increment by the gyroscopes will propagate the attitude in time. In the measurement update stage, the difference between the estimated and the measured Earth magnetic vector is implemented as feedback to correct the propagation error.

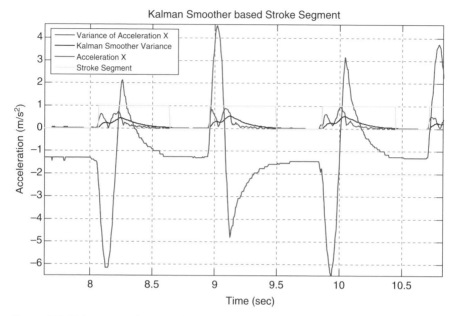

Figure 7.5. Kalman smoother-based stroke segment for motion status detection (2). See color insert.

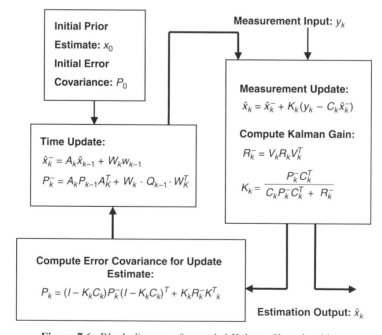

Figure 7.6. Block diagram of extended Kalman filter algorithm.

Figure 7.6 demonstrates the real-time recursive process of the extended Kalman filter algorithm.

7.3.2 Time Update Model

7.3.2.1 *Attitude Strapdown Theory for a Quaternion.* To propagate the attitude in time, the quaternion kinematics equation is:

$$\dot{q}(t) = \Omega(\omega_t)q(t) \tag{7.9}$$

where

$$\Omega(\omega_t) = \frac{1}{2} \begin{bmatrix} 0 & \omega_z & -\omega_y & \omega_x \\ -\omega_z & 0 & \omega_x & \omega_y \\ \omega_y & -\omega_x & 0 & \omega_z \\ -\omega_x & -\omega_y & -\omega_z & 0 \end{bmatrix} \tag{7.10}$$

$$q(t) = [\, q_1 \quad q_2 \quad q_3 \quad q_4 \,]^T \tag{7.11}$$

$$\omega_t = [\, \omega_x \quad \omega_y \quad \omega_z \,]^T \tag{7.12}$$

$q(t)$ is the quaternion that denotes the current attitude for the system state. ω_t are the current angular rates from the rate gyros for the system input. If Δt is small enough, the state matrix can be derived by the Euler method: $\Delta t = t - t_0 \rightarrow 0$,

$$q(t + \Delta t) \approx (I_4 + \Delta t \Omega)q(t) \triangleq f_{\Delta t}(q(t), \omega_{t+\Delta t}, \Delta t) \tag{7.13}$$

7.3.3 Error Model for Time Update

Equation 7.13 defines the nonlinear system propagation for the state q and input ω in time update. To obtain an extended Kalman filter with a capability of gyroscope bias separation, the sensor bias model is implemented in the sate matrix by error dynamics analysis.

We define the state error of a gyro as $\delta\omega$:

$$\omega_{\text{true}} = \omega_{\text{sensor}} - \delta\omega \tag{7.14}$$

$$\delta\omega(t) = \Delta\omega_{\text{bias}}(t) + w(t) \tag{7.15}$$

where $w(t)$ is a sensor's white noise and $\Delta\omega(t)$ is the gyro bias, which is considered as constant since dt is small:

$$\Delta\omega(t + \Delta t) = \Delta\omega(t) \tag{7.16}$$

The propagation of the attitude state error, δq can be obtained by partial differentiation of Equation 7.13:

$$\delta q(t + \Delta t) = \frac{\partial f_{\Delta t}(t)}{\partial \omega(t)} \delta \omega(t) + \frac{\partial f_{\Delta t}(t)}{\partial q(t)} \delta q(t) \qquad (7.17)$$

When time step dt and the previous δq is small, we assume $[\partial f_{\Delta t}(t)/\partial q(t)]\delta q(t) \approx 0$. By Jacobian linearization:

$$\frac{\partial f_{\Delta t}(t)}{\partial \omega(t)} = \frac{1}{2} \cdot \begin{bmatrix} q_4 & -q_3 & q_2 \\ q_3 & q_4 & -q_1 \\ -q_2 & q_1 & q_4 \\ -q_1 & -q_2 & -q_3 \end{bmatrix} \Delta t \triangleq -W_k \Delta t \qquad (7.18)$$

From Equations 7.13 and 7.17, the gyroscope bias can be separated from the system state:

$$q_k = f_{dt}(q_{k-1}, \omega_k, dt) - \delta q_k \qquad (7.19)$$

Thus, the discrete time update is

$$q_k = q_{k-1} + \Omega(\omega_k - \Delta \omega) \cdot q_{k-1} \cdot dt + W_k \delta \omega \qquad (7.20)$$

In the state-space representation,

$$x_{k+1} = A_k x_k + W_k w_k \qquad (7.21)$$

where

$$x_k = [q_k \quad \Delta \omega_k]^T \qquad (7.22)$$

$$A_k = \begin{bmatrix} [I_4 + \Omega(\omega_k - \Delta \omega_{k-1}) \cdot dt]_k & -\dfrac{\partial f_{\Delta t}(t)}{\partial \omega(t)} \\ 0 & I \end{bmatrix} \qquad (7.23)$$

7.4   MEASUREMENT UPDATE MODEL

After extended Kalman filter estimation, the spatial magnetic field disturbance becomes tolerable within one stroke and the Earth's magnetic field direction remains constant within the whiteboard. It can be used as a reference for attitude in the measurement update. The three orthogonal magnetometers in the MAG-μIMU measure the geomagnetic field with respect to (WRT) the body frame. On the other hand, by coordinate transform using the propagated attitude, it can be estimated from the constant geomagnetic field WRT the Earth frame. Hence the difference between the magnetometer measurements and the transformed geomagnetic field is feedback in the measurement update of the extended Kalman filter to correct for the error in attitude propagation.

Vectors $\vec{q}_b$, $\vec{q}_e$ are introduced to represent the geomagnetic field WRT the Earth frame and the magnetometer outputs, respectively. The two vectors are expanded into quaternions:

$$M_{\text{body}} = \begin{bmatrix} \vec{q}_b & 0 \end{bmatrix}^T, \qquad M_{\text{earth}} = \begin{bmatrix} \vec{q}_e & 0 \end{bmatrix}^T$$

If quaternion n denotes the current attitude, by coordinate transform:

$$M_{\text{body}} = n^* \otimes M_{\text{earth}} \otimes n \tag{7.24}$$

Multiplying the quaternion n to both sides of Equation 7.24, we obtain

$$n \otimes M_{\text{body}} - M_{\text{earth}} \otimes n = 0 \tag{7.24}$$

$$\begin{bmatrix} -[\vec{q}_b]_\times & \vec{q}_b \\ -\vec{q}_b^T & 0 \end{bmatrix} n - \begin{bmatrix} [\vec{q}_e]_\times & \vec{q}_e \\ -\vec{q}_e^T & 0 \end{bmatrix} n = 0 \tag{7.26}$$

where $[\vec{q}]_\times$ is the cross-product matrix:

$$[\vec{q}]_\times \triangleq \begin{bmatrix} 0 & -q_3 & q_2 \\ q_3 & 0 & -q_1 \\ -q_2 & q_1 & 0 \end{bmatrix} \tag{7.27}$$

From Equation 7.26:

$$\begin{bmatrix} -[(\vec{q}_b + \vec{q}_e)]_\times & \vec{q}_b - \vec{q}_e \\ -(\vec{q}_b - \vec{q}_e)^T & 0 \end{bmatrix} n = 0 \tag{7.28}$$

$$C \triangleq \frac{1}{2} \begin{bmatrix} -[(\vec{q}_b + \vec{q}_e)]_\times & \vec{q}_b - \vec{q}_e \\ -(\vec{q}_b - \vec{q}_e)^T & 0 \end{bmatrix} \tag{7.29}$$

Thus, there is no requirement for the state q to be a unit quaternion. Let C be the measurement matrix. The measurement update of the extended Kalman filter is

$$y = Cq \tag{7.30}$$

According to error dynamics analysis:

$$\delta q_k = q_k - q_k' \tag{7.31}$$

$$\delta y = \frac{\partial y}{\partial q_b} \delta q_b + \frac{\partial y}{\partial q_e} \delta q_e + \frac{\partial y}{\partial q} \delta q \tag{7.32}$$

where $[\partial y/\partial q_{be}]\delta q_{be} = 0$, $[\partial y/\partial q_b]\delta q_b \approx 0$ when the previous attitude state error is small.

From Equation 7.30:

$$\frac{\partial y}{\partial q_b} = \frac{1}{2} \begin{bmatrix} q_4 & -q_3 & q_2 \\ q_3 & q_4 & -q_1 \\ -q_2 & q_1 & q_4 \\ -q_1 & -q_2 & -q_3 \end{bmatrix} \tag{7.33}$$

Thus, the discrete measurement update is

$$y_k = C_k q_k - V_k \delta q_k \tag{7.34}$$

where

$$V_k = -\frac{\partial y}{\partial q_b} \tag{7.35}$$

A mathematical derivation method is introduced to derive an extended Kalman filter to minimize random noise and input bias error. The attitude calculation is totally based on a quaternion. As proved in Equation 7.28, the attitude quaternion q does not need to be unified in iteration. Furthermore, any reference field sensor, such as star sensors and accelerometers, or a combination, can be applied to this process model.

7.5   TESTING

7.5.1   Simulation Test

Extensive simulation experiments are performed to check the convergence of the MAG Extended Kalman filter. The simulation software includes two parts: the sensor output generation and the real-time filtering. To generate the sensor output, the digital pen's physical properties and motion are modeled by the mass, inertia matrix, input forces, and torques. The kinematics and dynamics module calculates the accelerations, angular rates, and magnetic field strength under ideal conditions. The sensors' outputs are synthesized by aliasing the random Gaussian noise and constant sensor bias. For the filter part, the sensor outputs of angular rates and magnetic field strengths are processed by the extended Kalman filter in real time. The attitude-in quaternion from the filter output is transformed into Euler angles for display.

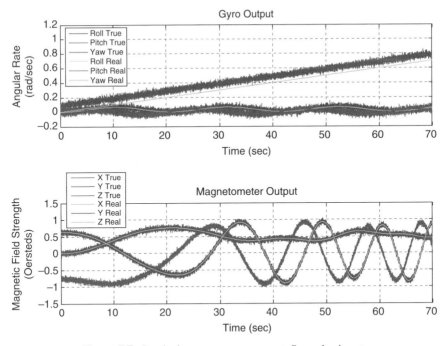

Figure 7.7. Synthetic sensor measurement. See color insert.

Figure 7.7 shows the simulated sensor output. For rotational motion analysis, the input forces are set as [0, 0, 0]. To simulate a complex orientation, the torque vector in roll, pitch, and yaw: [L, M, N] are set as

$$L = 0.01 \cdot \sin(0.3t) \tag{7.36}$$

$$M = 0.01 \cdot \cos(0.3t) \tag{7.37}$$

$$N = 0.01 \tag{7.38}$$

The zero mean Gaussian noises are added to the ideal sensor outputs. The absolute maximum error amplitudes are 1.3 degree per second for the gyros and 0.04 Oersteds for the magnetometers, respectfully. The initial attitude starts from 100 degrees in yaw angle. A constant sensor bias of 5 degrees per second is applied in yaw gyroscope output to verify the algorithm.

7.5.2 Experiment Test

Figure 7.8 illustrates the attitude result displayed in Euler angles. With the tracking ability of the extended Kalman filter, the initialization of the system state is simple.

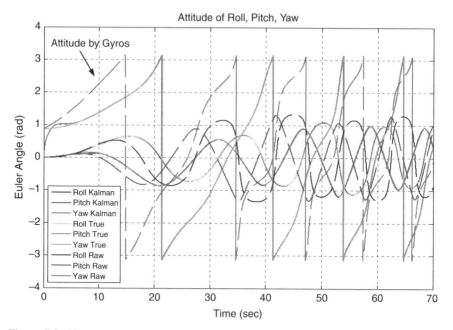

Figure 7.8. Simulation attitude comparison: filter result and gyroscope propagation. See color insert.

The attitude quaternion and gyro biases are set to zero. After iteration, the extended Kalman filter will estimate the gyro bias and remove it from the system state. According to magnetometer feedback, the filter's attitude estimation will converge and keep tracking automatically. The dashed line in Figure 7.8 shows the attitude propagated by the raw output from the gyroscopes. As shown in the figure, the random noise and bias error causes a large drift in the rolling, pitching, and yawing compared with the filter output.

The extended Kalman filter was tested using real sensor measurements. The MAG-μIMU transmits the digital sensor measurements to the computer wirelessly via the Bluetooth connection. The filter software in the computer processed the sensor data and calculated the attitude and sensor bias in real time.

The MAG-μIMU was held still for 4 seconds. Then continuous 90-degree rotations were performed to test the tracking performance. The sensor module was rotated counterclockwise for 90 degrees and clockwise back to 0 degrees along the sensing axis of the roll gyroscope. At the end of the seventh rotation, the MAG-μIMU was suddenly held still again to test the convergence capability from dynamic input to static input.

Figures 7.9 and Figure 7.10 show the six raw sensor output and the estimated attitude in Euler angles. Within the first iteration, the estimated attitude converged according to the observations from the magnetometers. The dashed lines show the attitude propagated by raw output from the gyroscopes. The sensor errors accumulated and caused the attitude drift.

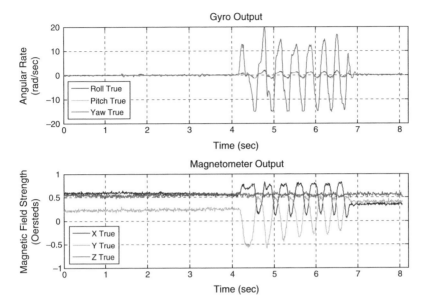

Figure 7.9. Real sensor data from MAG-μIMU. See color insert.

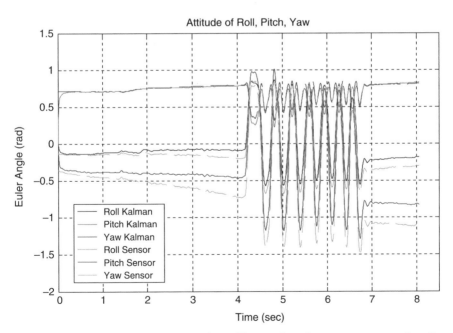

Figure 7.10. Experiment attitude comparison: filter result and gyroscope propagation. See color insert.

7.6 WRITING APPLICATION BASED ON ATTITUDE EKF COMPENSATION

The sensor measurements from MEMS accelerometers include gravitational accelerations as well as inertial accelerations. Because of the spring mass model of a sensor structure, these two terms cannot be separated from each other. The gravitational accelerations denote the constant gravity force projected to the sensing axes, which can be used as a reference for pen attitude. The inertial accelerations specify the translation and rotation of a rigid body when the pen is moving by handwriting. Hence, gravitational accelerations become a bias for inertial navigation, which needs to be balanced. On the other hand, during handwriting, the inertial acceleration interferes with the reference attitude calculation.

Magnetometers for attitude reference do not have such a disturbance because the magnetic field is irrelative to motion. However, the prevailing EMI will influence the measurement of the Earth's magnetic field in direction and strength. For instance, a conductor can bend the magnetic flux of the Earth's magnetic field because of magnetization. Besides, the Earth measurement may interfere with additive magnetic field error. As part of the electromagnetic field, the magnetic field exists when there is a changing electric field, and vice versa. A changing electric field can be caused by AC current, coil, capacitor, and antenna effect of conductor for electromagnetic induction. Another additive magnetic source may be directly induced by magnet, relay, monitors, transformer, and wireless communication device, such as mobile phone and Bluetooth. As a result, in real applications, the direction and strength of a measured magnetic field changes with different positions because of a complex environment.

Furthermore, the attitude from the accelerometer or magnetometer has ambiguity along the reference field direction, which is in the yaw and roll directions, respectively.

Assume the magnetic field noise is tolerable within one stroke space for the handwriting application; the tracking problem of the pen attitude is addressed by a complementary filter combining the ACC-EKF and MAG-EKF. In a time update, the attitude is propagated by angular rates based on the quaternion model. In a measurement update, during handwriting, the magnetometer model is implemented as an attitude reference with no inertial noise; and in the motionless stage between strokes, the estimation error caused by EMI can be rectified by accelerometers working as gravity tilter. The measurement update from the combination of ACC-EKF and MAG-EKF is complementary to filter the reference error and to compensate the attitude ambiguity for each other.

For complementary EKF, the measurement model of the attitude EKF, $y = Cq$, as introduced in Equation 7.30:

$$C \triangleq \frac{1}{2} \begin{bmatrix} -[(\vec{q}_b + \vec{q}_e)] \times & \vec{q}_b - \vec{q}_e \\ -(\vec{q}_b - \vec{q}_e)^T & 0 \end{bmatrix} \tag{7.39}$$

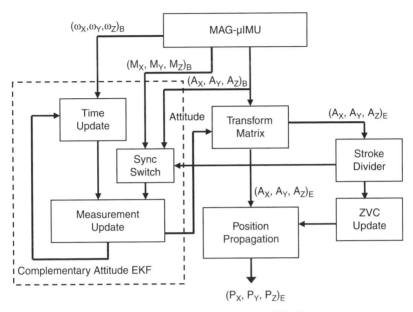

Figure 7.11. Complementary attitude EKF diagram.

where $\vec{q}_b$ is the measurement vector for the gravitational field or magnetic field as attitude reference. The $\vec{q}_e$ denotes the reference field vector in the Earth's frame. $\vec{q}_e = [0 \quad 0 \quad -1]^T$ is applied for accelerometer tilters and $\vec{q}_e = [1 \quad 0 \quad 0]^T$ is for magnetometers.

However, for continuous handwriting, the magnetic field varies in direction and strength corresponding to different places such that $\vec{q}_e \neq [1 \quad 0 \quad 0]$. For the complementary attitude EKF, before switching from accelerometer attitude feedback, the magnetometer measurement should be synchronized for the magnetic reference difference. From Equations 7.40 and 7.41, the measurement update model should be calibrated and initialized as

$$\vec{q}_e = \text{NORM}\left(R(\vec{q}) \cdot \vec{q}_b\right) \tag{7.40}$$

where

$$R(\vec{q}) = \begin{bmatrix} q_1^2 - q_2^2 - q_3^2 + q_4^2 & 2(q_1q_2 + q_3q_4) & 2(q_1q_3 - q_2q_4) \\ 2(q_1q_2 - q_3q_4) & -q_1^2 + q_2^2 - q_3^2 + q_4^2 & 2(q_2q_3 + q_1q_4) \\ 2(q_1q_3 + q_2q_4) & 2(q_2q_3 - q_1q_4) & -q_1^2 - q_2^2 + q_3^2 + q_4^2 \end{bmatrix} \tag{7.41}$$

The $\vec{q}$ is latest the attitude filter tracking result, and the $\vec{q}_b$ is the first magnetometer observation after the switch. The writing and pause of hand motion status for the Sync Switch can be implemented by the Stroke Segment Kalman Filter,

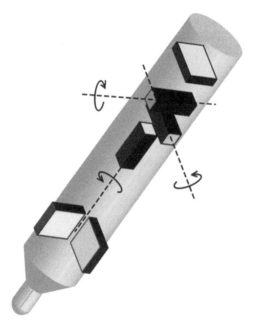

Figure 7.12. The MAG-μIMU system structure for a wireless digital writing instrument.

which will be introduced later. Figure 7.11 demonstrates the diagram of the complementary attitude EKF for the real-time handwriting application.

Figure 7.12 illustrates the MAG-μIMU sensor structure of the digital writing system for position tracking.

According to the strapdown kinematics theory [20] the body frame accelerations are transformed to the Earth's frame by a direct cosine matrix (DCM). After compensating for the gravitational and rotational accelerations, the translation accelerations integrate into 3D trajectories in space. Thus, any 2D human handwriting is recorded in real time if the pen touches the white board plane.

$$\vec{A}_b = \vec{A}_{\text{IMU}} - \vec{A}_{\text{Rotation}}(\omega_{\text{roll}}, \omega_{\text{pitch}}, \omega_{\text{pitch}}, L) \qquad (7.42)$$

$$\vec{V}_e = \text{DCM}(q)_b^e \vec{A}_b - \vec{G} \qquad (7.43)$$

$$\vec{P}_e = \int\int \vec{V}_e \qquad (7.44)$$

where $\vec{A}_{\text{IMU}}$ are the body frame accelerations: $[A_X, A_Y, A_Z]$. q is the quaternion representing the pen attitude. $\vec{G}$ is the gravity vector: $[0, 0, -g]$.

According to the kinematics equations, the accelerometers are designed to mount close to the pen-tip for more sensitivity for handwriting motion and depression for

Figure 7.13. Experiment setup for Digital Writing Instrument with MAG-μIMU.

the rotational accelerations. The magnetometers are fixed on the pen bottom to reduce the magnetic field distortion effect by the metal white board.

7.7 EXPERIMENTAL RESULTS OF AN INTEGRATED SYSTEM

The handwriting experiments are performed to verify the algorithm integration for the Digital Writing Instrument system and to test the performance of attitude and position tracking for handwriting recording.

Figure 7.13 illustrates the writing experiment setup with the wireless Digital Writing Instrument based on MAG-μIMU.

In this experiment, an 8-cm × 10-cm capital letter "A" with four strokes is written on a horizontal plastic table in normal writing speed. During handwriting, the MAG-μIMU measured the nine-channel motion information of X, Y, Z acceleration, roll, pitch, yaw angular rate, and X, Y, Z magnetic field strength with a 200-Hz sampling rate. After wireless transmission via Bluetooth, the complementary attitude EKF in a computer estimated the pen attitude in real time with a dynamic switch to combine the accelerometer and the magnetometer update, which is the controlled by stroke segment Kalman filter. After coordinate transform by attitude tracking result, and zero velocity compensation (ZVC), the handwriting trajectory in the Earth's frame was obtained stroke by stroke. Figure 7.14 shows the nine-channel sensor output WRT the body frame from the MAG-μIMU.

The motion information for 3D accelerations, angular rates, and magnetic field strength WRT the body frame are transformed after calibration experiments as shown in Figure 7.15 Random noise prevails in sensor observations.

Meanwhile, the body frame acceleration in the x-axis is used to detect the writing or pause of hand motion by stroke segment Kalman filter in real time (Figure 7.16). This writing status information can control the switching in the measurement update.

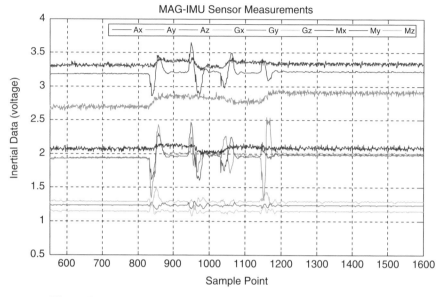

Figure 7.14. Digital Writing Instrument sensor outputs. See color insert.

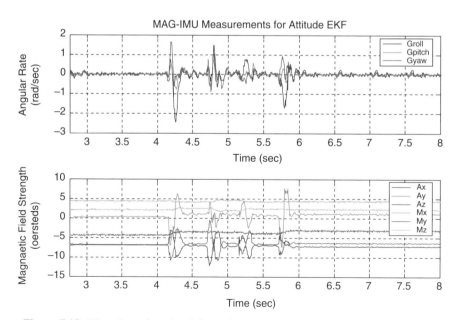

Figure 7.15. Nine-channel motion information WRT the sensor frame. See color insert.

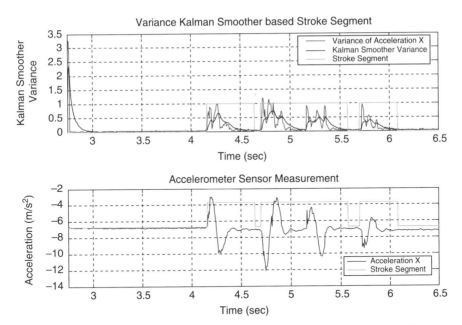

Figure 7.16. Stroke Segment Kalman Filter as sync switch for measurement update. See color insert.

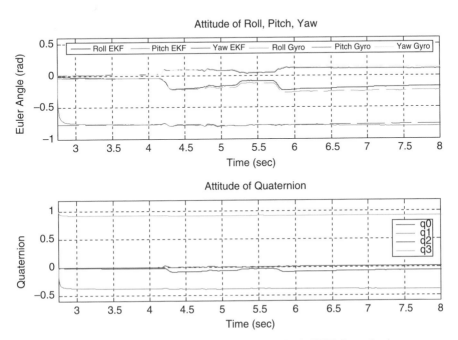

Figure 7.17. Tracking result of complementary attitude EKF. See color insert.

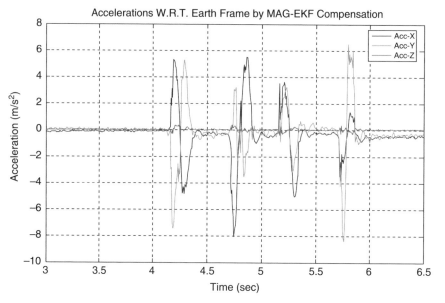

Figure 7.18. Accelerations WRT the Earth's frame by complementary attitude EKF. See color insert.

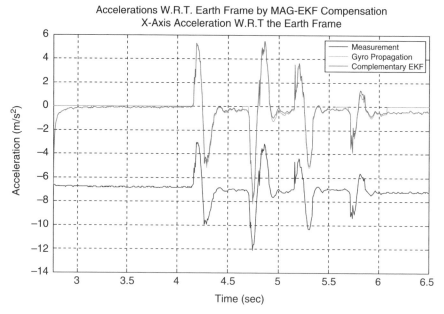

Figure 7.19. Performance comparison for the acceleration coordinate transform in x-axis. See color insert.

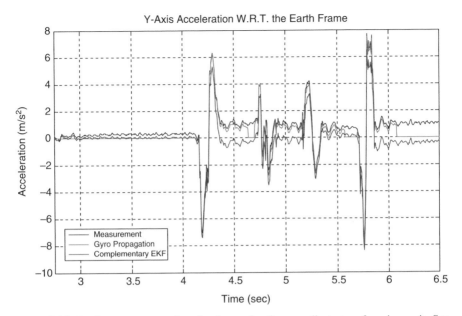

Figure 7.20. Performance comparison for the acceleration coordinate transform in y-axis. See color insert.

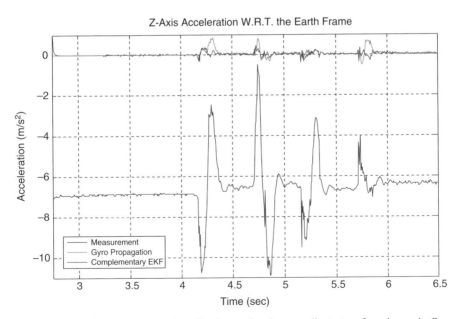

Figure 7.21. Performance comparison for the acceleration coordinate transform in z-axis. See color insert.

The pen attitude in quaternion is estimated by the complementary attitude EKF in real time. For comparison, the attitude is also displayed in Euler angles as shown in Figure 7.16. After 0.26 sec, the filter is tracked to true attitude for 44.3 degrees in roll.

As shown in Figure 7.17, the acceleration WRT the Earth's frame is obtained by coordinate transform according to the filter result for pen attitude in real time.

Figure 7.18, 7.19–7.21 demonstrate the performance for the transformation of the acceleration's coordinate from the sensor frame to the writing frame by the complementary attitude EKF quaternion output. For comparison, the transformed acceleration by noisy gyroscope propagation and the sensor measurement are shown on the green and blue lines, respectively.

During handwriting, the pen attitude is changing with hand motion. The attitude error caused by noisy measurement will lead to nonlinear bias by coordinate transform for the Earth's frame accelerations.

As the pen tip is moving on the write board which is the $x-y$ plane of the Earth's coordinate system, the Earth's frame acceleration on the z-axis should equal zero.

The real-time writing result for "A" is shown in Figure 7.22, which started from the origin (0,0). Because of noise and bias in the accelerometer measurements, the position error was accumulated and magnified by double integration and caused it nonlinear drift and distortion in the writing trajectory.

Thus, the zero velocity compensation is implemented to improve position accuracy. The stroke segment is improved using the z-axis acceleration in Earth's frame by coordinate transform and the complementary EKF tracking result (Figure 7.23).

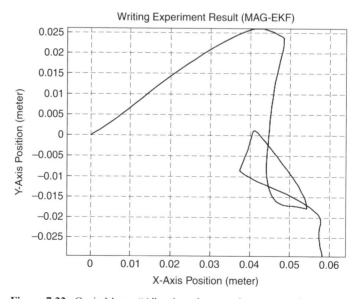

Figure 7.22. Capital letter "A" written by complementary attitude EKF.

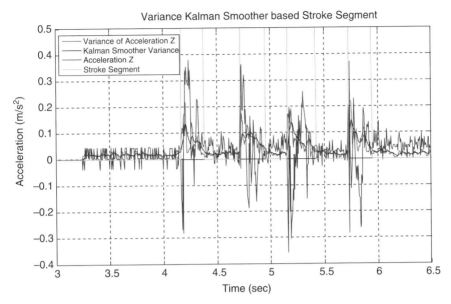

Figure 7.23. Stroke segment according to the Earth frame acceleration. See color insert.

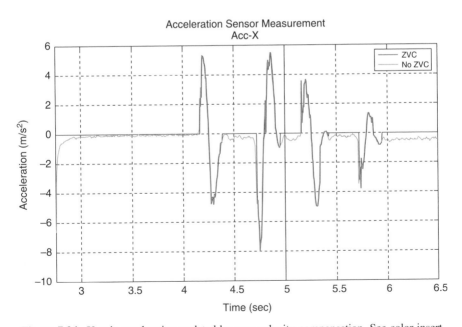

Figure 7.24. X-axis acceleration updated by zero velocity compensation. See color insert.

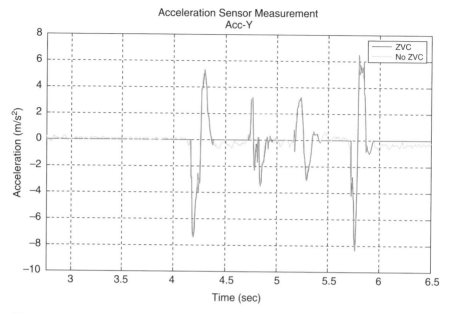

Figure 7.25. Y-axis acceleration updated by zero velocity compensation. See color insert.

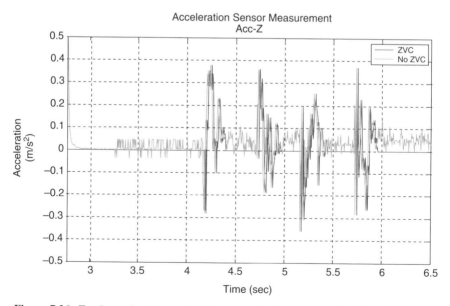

Figure 7.26. Z-axis acceleration updated by zero velocity compensation. See color insert.

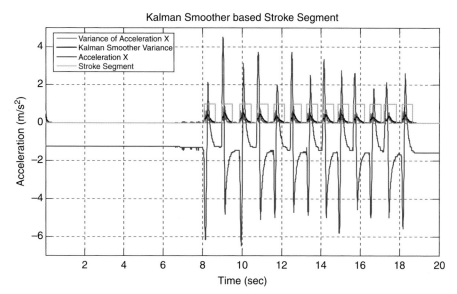

Figure 7.4. Kalman smoother-based stroke segment for motion status detection (1).

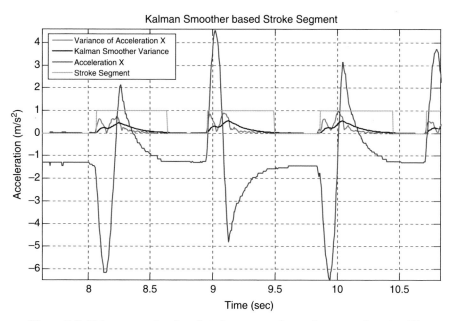

Figure 7.5. Kalman smoother based stroke segment for motion status detection (2).

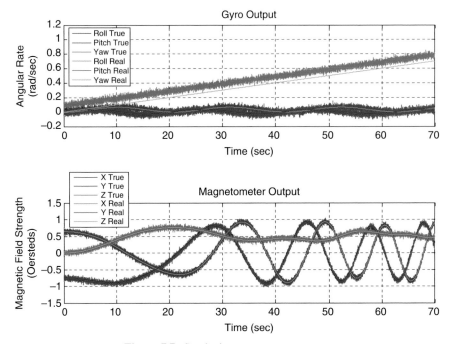

Figure 7.7. Synthetic sensor measurement.

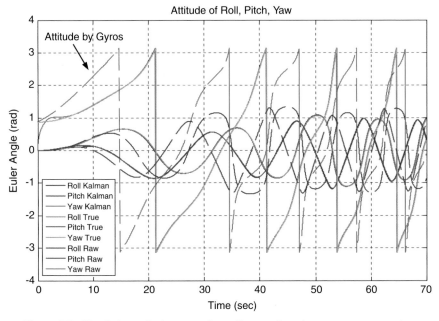

Figure 7.8. Simulation attitude comparison: filter result and gyroscope propagation.

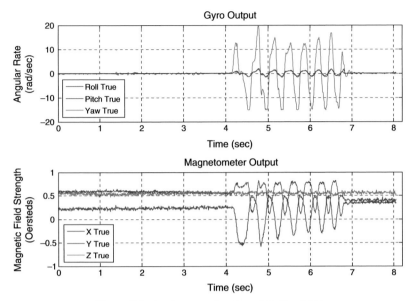

Figure 7.9. Real sensor data from MAG-μIMU.

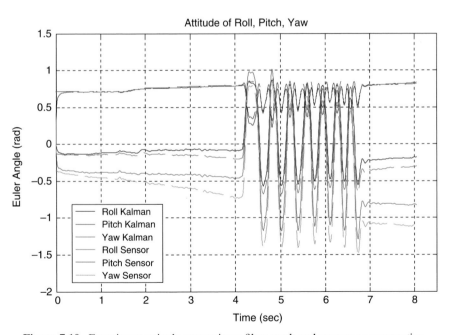

Figure 7.10. Experiment attitude comparison: filter result and gyroscope propagation.

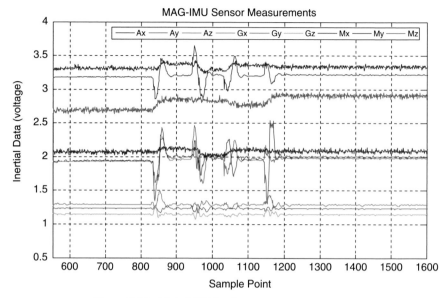

Figure 7.14. Digital Writing Instrument sensor outputs.

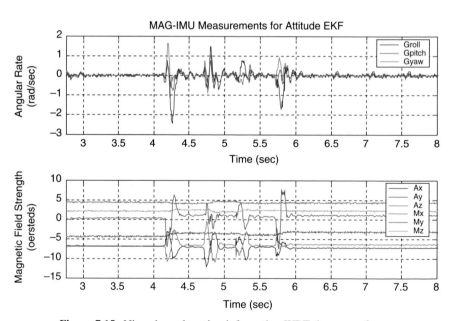

Figure 7.15. Nine-channel motion information WRT the sensor frame.

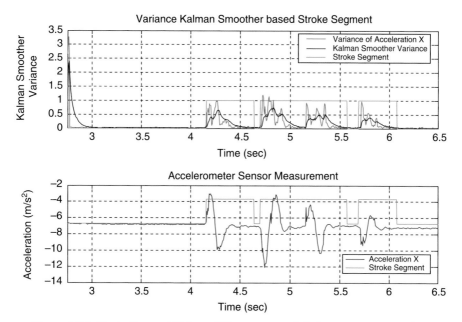

Figure 7.16. Stroke Segment Kalman Filter as sync switch for measurement update.

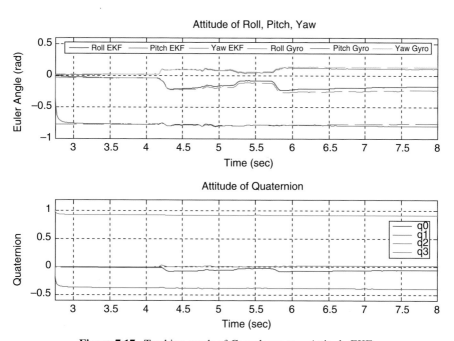

Figure 7.17. Tracking result of Complementary Attitude EKF.

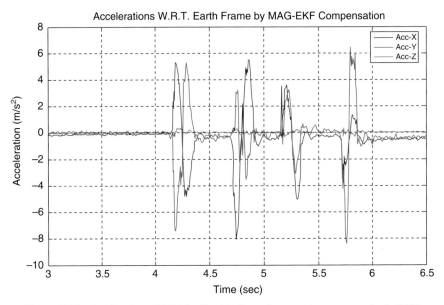

Figure 7.18. Accelerations WRT the Earth's frame by complementary attitude EKF.

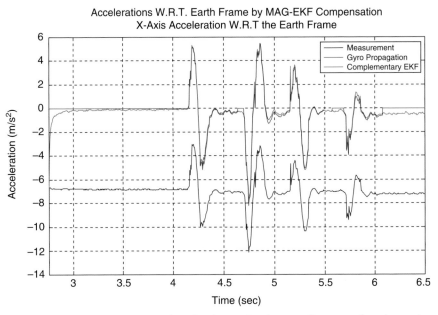

Figure 7.19. Performance comparison for the acceleration coordinate transform in x-axis.

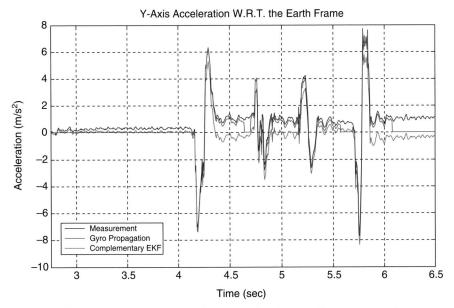

Figure 7.20. Performance comparison for the acceleration coordinate transform in y-axis.

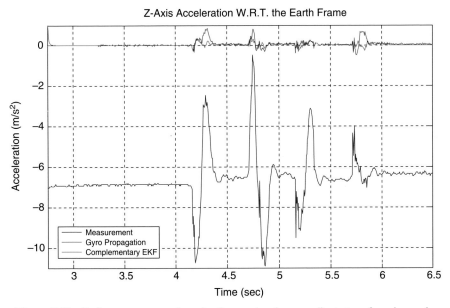

Figure 7.21. Performance comparison for the acceleration coordinate transform in z-axis.

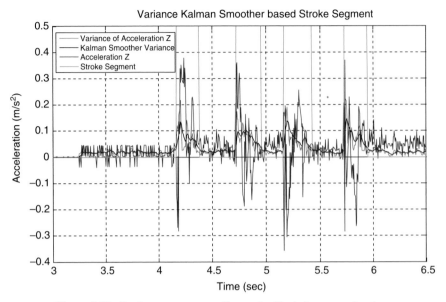

Figure 7.23. Stroke segment according to the Earth frame acceleration.

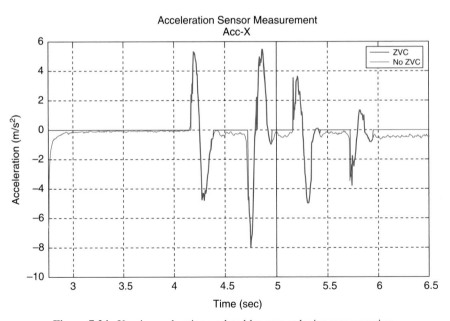

Figure 7.24. X-axis acceleration updated by zero velocity compensation.

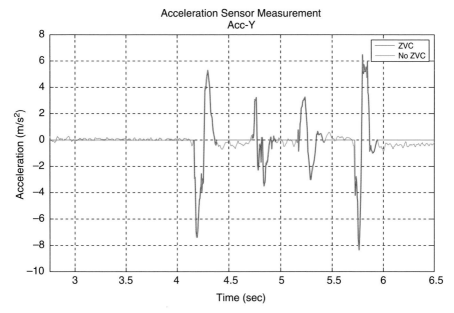

Figure 7.25. Y-axis acceleration updated by zero velocity compensation.

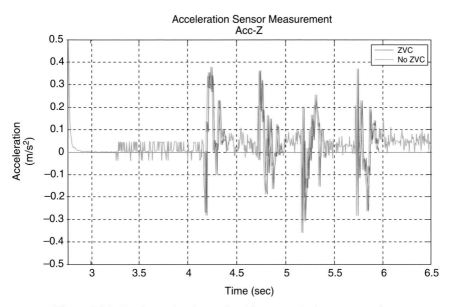

Figure 7.26. Z-axis acceleration updated by zero velocity compensation.

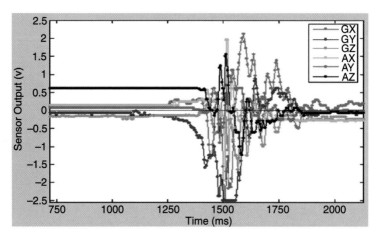

Figure 8.12. Original motion data recording of falling-down.

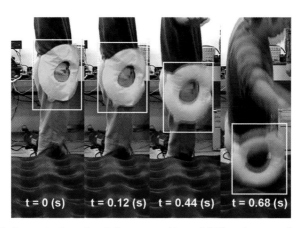

Figure 8.13. A demonstration of real-time recognition of falling-down motion with inflation of airbag.

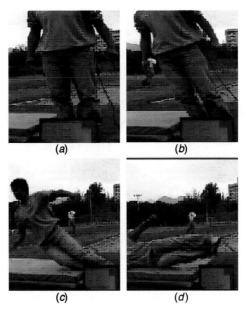

Figure 8.16. Lateral falling-down state.

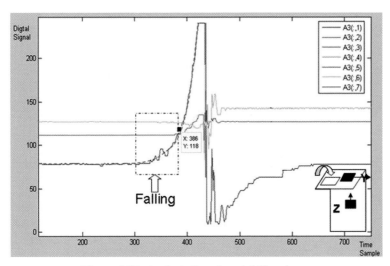

Figure 8.17. Analysis for a lateral falling-down.

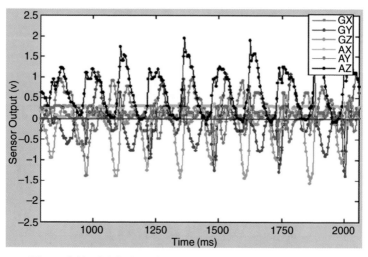

Figure 8.19. Original motion data recording of forward running.

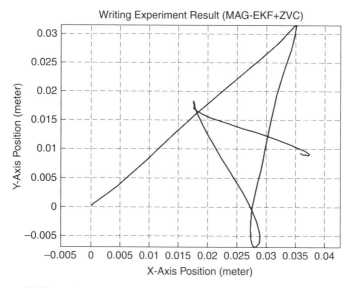

Figure 7.27. Writing result for "A" by complementary attitude EKF and ZVC.

Figure 7.28. Handwriting result for "A" using DWI system during the position tracking experiment.

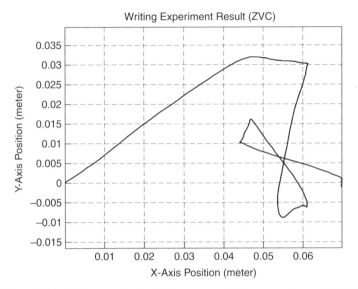

Figure 7.29. Writing result for "A" by gyro attitude propagation and ZVC.

Figures 7.24–7.26 illustrate the acceleration updated by ZVC. During the pause stage of writing, the velocity was set to zero and the position remained constant to avoid position drift.

Velocity can be corrected by ZVC. The error accumulated in velocity calculation from accelerometer noise and bias, and magnetic field distortion was rectified backward after each stroke. When the sampling rate is high enough and magnetic field distortion within one stroke is tolerable, the higher order term can be omitted and the velocity error can be approximated in the linear model. However, ZVC is an offline algorithm. The letter should be written stroke by stroke to use velocity bias.

After coordinate transform using a complementary EKF attitude and ZVC, the handwriting trajectory for "A" is improved as shown in Figure 7.27.

Figure 7.28 demonstrates the handwriting result during this experiment for comparison. The capitialized "A" is written by the marker in the Digital Writing Instrument on A4 size paper.

Compared with Figure 7.25, there is drift error and distortion in Figure 7.29, which is obtained by coordinate transform using gyro attitude propagation and ZVC.

7.8 CONCLUSION

A wireless Digital Writing Instrument based on MAG-μIMU has been developed with 3D accelerometers, gyroscopes, and magnetometers with strapdown installment. Through a Bluetooth connection, the motion data are transmitted to a computer with a 200-Hz sampling rate for real-time process in attitude determination and position tracking.

REFERENCES

1. Virtual Ink Corporation. Mimio Studio user's guidance, 2005. http://www.mimio.com/products/documentation/mimiostudio_usersguide.pdf.

2. Lucia, Inc. eBeam Interactive Whiteboard Technology, 2006. http://www.e-beam.com/downloads/files/WhatIseBeam.pdf.

3. Sagem Avionics, Inc. APIRS F200-AHRS Brochure, 2006. http://www.sagemavionics.com/Products/Brochures/APIRS-F200.pdf.

4. Watson Industries, Inc. AHRS-E304 Manual, 2006. http://www.watson-gyro.com/files/attitude_reference_AHRS-E304_spec.pdf.

5. R. C. Hayward, D. Gebre-Egziabher, M. Schwall, J. D. Powell, and J. Wilson, "Inertially-aided GPS-based attitude heading reference system (AHRS) for general aviation," in *Proc. of ION GPS*, Kansas City, MO, Sept. 1997.

6. A. K. Brown, "Test results of a GPS/Inertial navigation system using a low cost MEMS IMU," in *Proc. 11th Annual Saint Petersburg International Conference on Integrated Navigation System*, Saint Petersburg, Russia, May 2004.

7. E.-S. Choi, W. Chang, W. C. Bang, J. Yang, S. J. Cho, J. K. Cho, J. K. Oh , and D. Y. Kim, "Development of the gyro-free handwriting input device based on inertial navigation system (INS) theory," in *Proc. of The Society of Instrument and Control Engineers Annual Conference*, Vol. 2, 2004, pp. 1176–1181.

8. G. Zhang, G. Y. Shi, Y. L. Luo, H. Wong, Wen J. Li, Philip H. W. Leong, and M. Wong, "Towards an ubiquitous wireless digital writing instrument using MEMS motion sensing technology," in *IEEE/ASME Proc. Advanced Intelligent Mechatronics*, 2005, pp. 795–800.

9. Y. Luo, C. C. Tsang, G. Zhang, Z. Dong, W. J. Li, and P. H. W. Leong, "An attitude compensation technique for a MEMS motion sensor based digital writing instrument,"in *Proc. of IEEE Int. Conf. on Nano/Micro Engineered and Molecular Systems*, 2006.

10. Analog Devices, Inc. Low-Cost Ultracompact $\pm 2g$ Dual-Axis Accelerometer ADXL311 Data Sheet (Rev. B, 01/2005), 2005. http://www.analog.com/UploadedFiles/Data_Sheets/243920868ADXL311_B.pdf.

11. Murata Manufacturing Co., Ltd. GYROSTAR Piezoelectric Vibrating Gyroscopes Data Sheet (No. S42E, 06/2006), 2006. http://www.murata.com/catalog/s42e.pdf.

12. Koninklijke Philips Electronics N. V. Magnetic Field Sensor KMZ51 Data Sheet (06/2000), 2006. http://www.nxp.com/acrobat_download/datasheets/KMZ51_3.pdf.

13. Koninklijke Philips Electronics N. V. Magnetic Field Sensor KMZ52 Data Sheet (06/2000), 2006. http://www.nxp.com/acrobat_download/datasheets/KMZ52_1.pdf.

14. Honeywell International, Set/Reset Function for Magnetic Sensors (08/02), 2003. http://www.ssec.honeywell.com/magnetic/datasheets/an213.pdf.

15. Atmel Corporation. 8-bit AVR Microcontroller with 32 Bytes In-System Programmable Flash ATmega32(L) Data Sheet (Rev. I, 04/06), 2006. http://www.atmel.com/dyn/resources/prod_documents/doc2503.pdf.

16. R. H. Barnett, S. Cox, and L. O'Cull, *Embedded C Programming and the Atmel AVR*, Delmar, Clifton Park, NY, 2003.

17. BlueRadios, Inc. Class 1, Class 2 and 3 BR-C30 ver 1.2 Module Specification, 2006. http://www.blueradios.com/BR-C30.pdf.

18. IVT Corporation, BlueSoleil 2.3 Commercial Version Release Note (Ver. 2.3, 07/06), 2006. http://www.bluesoleil.com/Download/files/IVT_BlueSoleil_%28Standard%29_Release_Note.pdf.

19. Future Technology Devices International Ltd. FT232BM USB UART (USB-Serial) I.C. Data Sheet (Ver. 1.8, 2005), 2006. http://www.ftdichip.com/Documents/DataSheets/ds232b18.pdf.

20. D. H. Titterton and J. L. Weston, *Strapdown Inertial Navigation Technology*, 2nd Edition, AIAA, Reston, VA, 2005.

FURTHER READING

W.-C. Bang, W. Chang, K.-H. Kang, E.-S. Choi, A. Potanin, and D.-Y. Kim, "Self-contained spatial input device for wearable computers," in *Proc. 7th IEEE Int. Symp. on Wearable Computers*, pp. 26–34, Oct. 2005.

D. Gebre-Egziabher and D. Powell "A DME-based area navigation systems for GPS/WAAS interference mitigation in general aviation applications," in *Proc. IEEE Position Location and Navigation Symposium*, pp. 74–81, Mar., 2000.

L. Monteiro, T. Moore, and C. Hill, "What is the accuracy of DGPS?" *The Journal of Navigation*, Vol. 58, pp. 207–225, 2005.

Wikipedia foundation Inc. Global Positioning System, 2006. http://en.wikipedia.org/wiki/Global_Positioning_System.

CHAPTER 8

Intelligent Mobile Human Airbag System

8.1 INTRODUCTION

A human airbag system is designed to reduce the impact force from slippage falling-down. A micro inertial measurement unit (μIMU), which is based on Micro ElectroMechanical System (MEMS) accelerometers and gyro sensors is developed as the motion sensing part of the system. A recognition algorithm is used for real-time falling determination. With the algorithm, the microcontroller integrated with μIMU can discriminate falling-down motion from normal human motions and trigger an airbag system when a fall occurs. The airbag system is designed to be a fast response with moderate input pressure; i.e., the experimental response time is less than 0.3 sec. under 0.4 MPa. Also, we present our progress on using this μIMU based on support vector machine (SVM) training to recognize falling-motions. Experimental results show that selected eigenvector sets generated from 200 experimental data can be separated into falling-motion and other motions completely.

It is well known that the world is facing an increasingly aging population. With this increase, the proportion of frail and dependent elderly is also likely to increase significantly [1]. This shift in demographic pattern will lead to an exponential increase in the number of elder individuals suffering from injury of falls; i.e., falls and fall-induced fractures are very common among the elderly.

Hip fractures account for most of the deaths and costs of all fall-induced fractures. Apart from causing physical injury, hip fracture can result in significant psychologic trauma and lead to self-imposed restrictions of activity that can compromise the quality of life of the individual [2]. Hip protectors are protective devices made of hard plastic or soft foam and are placed over the greater trochanter of each hip to absorb or shunt away the energy during mechanical impact on the greater trochanter. They are widely demonstrated both biomechanically and clinically to be capable of reducing the incidence of hip fractures. However, the compliance of the elderly to wear them is very low, because of discomfort, wearing difficulties, problems with

Intelligent Wearable Interfaces, by Yangsheng Xu, Wen J. Li, and Ka Keung C. Lee
Copyright © 2008 John Wiley & Sons, Inc.

urinary incontinence and illness, physical difficulties, and them being not useful and irrelevant. Our group is developing a novel hip protector with smaller dimensions and greater comfort for the elderly. Basically, a MEMS motion sensing unit will be used to detect imbalance and to trigger the inflation of compact airbags worn by the elderly.

Two key issues have to be considered for the human airbag system. One is a comfortable compressed airbag that can be inflated instantly. Another is a small triggering device embedded with a rapid and accurate algorithm for recognizing falling motion. For the fast inflated airbag, we first use an electromagnetism valve as a switch to open a compressed air source. When the valve is triggered, an airbag can be inflated in less than 0.3 sec. under 0.4 Mpa, which is enough to effectively reduce the impact force on the hip caused by a fall. Now, an independent airbag is under development based on compressed CO_2 cartridges. In this project, we will focus our discussion on the triggering device and on the motion-recognition algorithm.

Because of the availability of low-cost, small-size MEMS sensors, it is possible to build self-contained inertial sensors with an overall system dimension of less than 1 cubic inch, and at the same time, the sensing unit can track the orientation and other motions in real time. As an example, our group developed the micro input devices system (MIDS) based on MEMS sensors as a novel multifunctional interface input system, which could potentially replace the mouse, pen and keyboard as input devices to the computer [3, 4]. We also developed a μIMU, which measures three-dimensional angular rates and accelerations based on MEMS sensors. This system is similar to the MIDS but has a different hardware configuration and uses different software protocols. A USB port was designed for data transmission on the μIMU. With the μIMU, we recorded human motion, including normal motions and falling. An algorithm SVM was used to analyze the data recorded by the small MEMS unit.

Recently, we integrated a microcontroller and Bluetooth module on the μIMU, and the overall size of the unit is designed to be less than 26 mm × 20 mm × 20 mm. The μIMU is an essential part of the novel hip protector, which can collect human motion data wirelessly and recognize motion data, e.g., falling-motion using recognition algorithm. A mechanism of triggering the airbag is also developed, and we proved that triggering an airbag system by the μIMU is feasible.

Generally, most complicated pattern recognition problems involve the identification of dynamic and time-varying signals, such as speech recognition, handwriting recognition, and image sequences identification. Signals of MEMS sensor output concerning daily physical activities are of low frequency, transient, and dynamic in nature. In Ref. 1, an eigenvector-based pattern recognition method was used to initiate multidimensional signal identification to analyze MEMS accelerometer's data. However, the sensing unit did not use gyro sensors, and consequently, a lot of rotation information was not used. A better classification result for recognition of different human motions was reported by Mantyjarvi et al. [5] who used independent component analysis and principal component analysis, and they achieved 83% to 90% accuracy. We present our progress on using our μIMU with SVM training to recognize falling-motions in this project. Experimental results show that selected eigenvector sets generated from 200 experimental data can be separated into falling and other motions successfully.

8.2 HARDWARE DESIGN

As mentioned, hip fractures account for most of the costs of falls and fall-induced fractures, especially for elderly people. We propose to develop intelligent and personalized wearable airbags to reduce the force of impact during a fall for the elderly. Recent advances in manufacturing technologies have made it possible to safely compress air in small, light-weight, and low-cost pressurized cylinders, thereby making a personalized airbag system not only tractable, but also economically feasible. In addition, a MEMS-based inertia measurement unit is suitable for a small, light-weight hip protector system, and it can be intelligently programmed to measure and recognize human motions to trigger the inflation of the airbag(s) before a subject falls to the ground.

Figure 8.1 illustrates the basic concept of an intelligent human airbag system. Initially, the airbag is compressed in a belt. When an elderly person loses balance, the MEMS micro sensors in the belt will detect his/her disorientation and trigger the inflation of the air bag on the side in a few milliseconds before falling to the ground. The hip-airbags can be designed just like automobile airbags, which

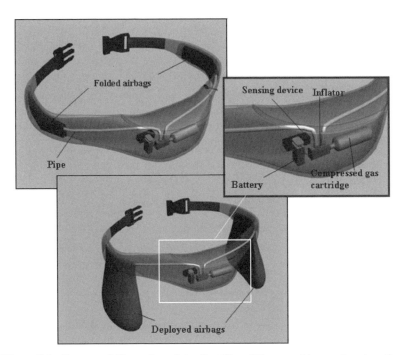

Figure 8.1. Conceptual illustration of the "intelligent" human airbag system in action.

contain many micron-size holes for automatic deflation. Therefore, distension can be controlled to last for a few seconds, and the hip-airbags will gradually collapse afterward. The force attenuation property of the inflated hip protector will be tested using the established method in our laboratory. The motion-based condition of activating the inflation process will be defined such that it is sensitive enough to detect the imbalance of an elderly person but not hypersensitive and induce false alarms. Testing for a falling-down condition and generating a trigger signal via the μIMU is the key issue discussed in this project.

8.2.1 μIMU System Design

We use the ATMEL ATmega32 microcontroller in the design with eight channels of 10-bit ADC, an USART (universal synchronous and asynchronous serial receiver and transmitter) port, and so on [6]. The Bluetooth module is connected with the microcontroller by the UART at a baud rate of 56.7 KHz.

For our experiments, we use ADXL203 and ADXRS300 sensors as accelerometers [7] and angular rate gyros [8], respectively. These sensors are produced by analog devices, and they are low-cost and relatively high-performance sensors with analog signal output. The output signals of the accelerometers (a_x, a_y, a_z) and the rate gyros (ω_x, ω_y, ω_z) are measured directly with an A/D converter inside the microcontroller. The digital sample rate of the microcontroller is 200 Hz, which ensures rapid reaction to human motion. Three gyroscope sensors (single-axis) and two accelerometers (dual-axes) were used in our system.

The accelerometers and the gyros act as an μIMU of the motion sensing system. These μIMU sensors and the Bluetooth module are housed on a small printed circuit board (PCB), as shown in Figure 8.2.

We adopt a Bluetooth Module in our system to transfer data to a host system [6]. This module provides easy integration to various host systems. The module is directly connected to the microcontroller via a USART port. The module is very small in size (69 mm × 24 mm × 5 mm) and can easily communicate with the microcontroller. Thus, the μIMU can realize two functions:

1. Data collection and transmission to the computer wirelessly, which can be analyzed or trained using a SVM.
2. A recognition algorithm can be downloaded to discriminate a falling motion and to trigger the airbag for inflation.

8.2.2 Mechanical Release Mechanism

Two mechanisms were tried as the inflator. The first mechanism uses a solenoid valve to trap the gas released from a punctured pressurized cylinder. A commercially available inflator that has a fixed puncture pin inside is connected to the inlet of a two-way solenoid valve SLG5404-04 (EMC Co. Ltd., Ningbo, China). Then the pressurized cylinder is screwed to and punctured by the inflator at the same time. The released

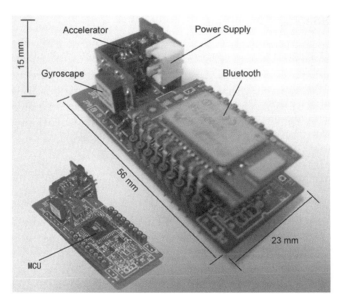

Figure 8.2. Photograph of a 3D motion sensing system consisting of three gyros and three acceleration sensors.

gas is trapped by the valve. The valve opens and deploys the airbags at the outlet once it is triggered by the μIMU.

Another one is punching the pressurized cylinder only when a fall is detected. The mechanism as shown in Figure 8.3, includes a punch mounted on a launcher that consists of a spring and a locking switch. The spring is compressed by screwing. When the locking switch is pressed by the actuator, the compressed spring extends and the punch accelerates toward the pressurized cylinder. The pressurized cylinder is allowed to be pushed back by the jet so that the punched hole will not be blocked by the punch.

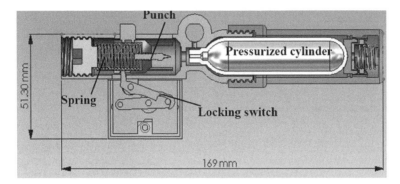

Figure 8.3. The cross-sectional view of the second mechanism of the inflator.

8.2.3 Minimization of Airbag Inflation Time

We have designed an experiment to test how the input pressure affects the inflation time of the inflation system.

As shown in Figure 8.4, we used a laboratory compressed air source with different pressures for our initial experiments. The solenoid valve is used as a switch for the compressed air. The valve has an orifice of 1.2 mm and a response time as short as 15 ms. Compressed air was first trapped inside a pipe of about 131 mL connected to the inlet of the valve under different pressures. In addition, we used a relay to deliver 12 V from a power supply to the solenoid valve.

We modulated the compressed air source from 1 bar to 6 bars (gage pressures) and used 5 V as a switch voltage to open or close the solenoid valve since the micro control unit (MCU) signal to indicate a fall will be 5 V in our final system design.

Figure 8.5 shows the inflation time of the airbag under different inlet pressures. As shown, for any inlet pressure greater than two bars, inflation time is less than 0.5 sec. In particular, when the gage pressure is four to six bars, the inflation time can be

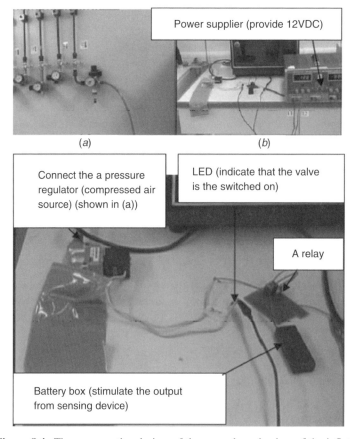

Figure 8.4. The cross-sectional view of the second mechanism of the inflator.

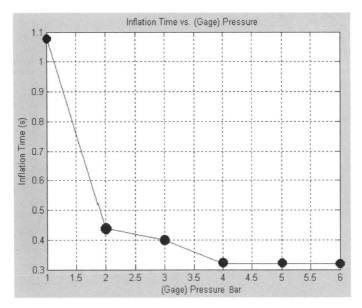

Figure 8.5. Results of continuous inlet pressure VS inflation time.

limited to 0.3 sec, which ensures the airbag can be deployed before a human falls to the ground. From our own laboratory high-speed camera image analysis, a typical fall takes 0.4–0.5 sec from the beginning to the end, and hence, a pressure of four bars should be sufficient to inflate an airbag to reduce the impact force on a person from a fall.

Although the valve has a quick response, there is a serious leakage problem when connecting to the pressurized cylinder. Furthermore, the size of valve has to be large in order to sustain the high pressure of the cylinder. Thus, we abandoned this mechanism and switched to the second mechanism as described below.

8.2.4   The Punch Test for the Second Mechanism

The punching mechanism can be simply represented by Figure 8.6. Three parameters are varied to create a hole on the cap of the cylinder as large as possible. The parameters are the compression of the spring e_0 (or the spring force), the angle of the punch end $\dot{\mathrm{E}}$, and the traveling distance of punch D0. The traveling distance of the punch is also important because impact force is larger as the rate of loading increases. A nail of square pyramidal tip of different angles $16°$ and $19°$ (mm) were used to punch Gamo (an air gun manufacturer) 12-g pressurized cylinders. A spring of spring constant 1514 N/m was used. Square holes of different sizes were created and listed as in Table 8.1. Figure 8.7 shows the discharging time (time taken for pressure inside to drop to atmospheric pressure) of 12-g pressurized cylinders against the spring force inputted.

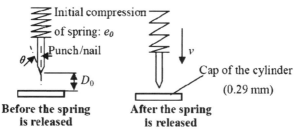

Figure 8.6. The setup of the second mechanism.

Table 8.1. The Results of a Square Cross-Nail to Punch Pressurized Cylinders

	θ	e_0 (mm)	D_0 (mm)	Length of Punched Hole (mm)
1	16°	40.69	15.12	0.42
2	16°	50.57	15.12	1.13
3	16°	42.20	10	0.98
4	16°	54.20	10	1.2
5	19°	40.67	14.3	0.65
6	19°	50.57	14.3	0.73

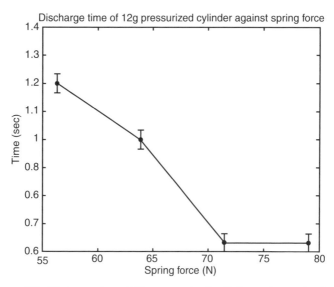

Figure 8.7. Discharge time of 12-g pressurized cylinder against spring force.

The cross-sectional area of the hole created by the nail can be as large as $1.2 \times 1.2\,mm^2$, which is greater than the orifice of the solenoid valve. Moreover, the discharge coefficient is proportional to:

a) The ratio of the orifice diameter to the pipe's diameter.
b) The Reynolds Number.

The ratios of diameters for the valve and the second design are 0.4 and 0.21, respectively, and the Reynolds Numbers are of order 104 and 106, respectively. From the plot of the orifice discharge coefficient against the Reynolds Number, the discharge coefficients of the valve and the second design are different by less than 0.015.

By combining the charging and discharging models of updating the upstream and downstream pressures sequentially, we estimate the size of the punched hole required to inflate a 1200-mL airbag within 0.3 sec. Consider the 531-mL airbag connected to the valve as a 531-mL vacuum bottle, and take the discharge coefficient as 0.6 and the specific heat ratio as 1.4 for air. Figure 8.8 shows estimation of the pressures inside the gas source and the airbag. The estimated inflation time is 0.38 sec, which is closed to the measured value. Figure 8.9 shows the estimation for a 16-g CO_2 pressurized cylinder with a square punched hole of length 1.2 mm. The result shows that the 16-g pressurized cylinder can inflate the 1200-mL airbag to 0.28 MPa within 0.3 sec. Therefore, it is possible for the second design to inflate the same airbag in 0.4 sec (including the response time of the actuator).

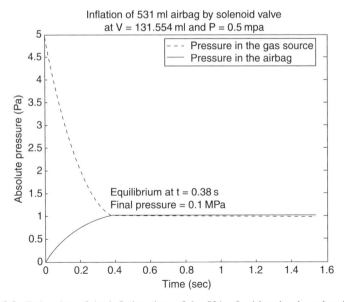

Figure 8.8. Estimation of the inflation time of the 531-mL airbag by the solenoid valve.

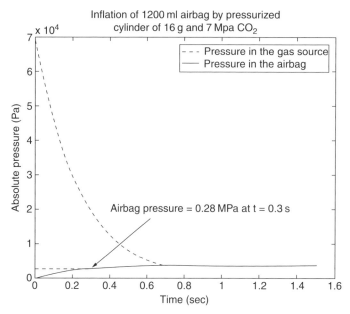

Figure 8.9. Estimation of the inflation time of the 1200-mL airbag by a 16-*g* pressurized cylinder.

8.2.5   System Integration

The motion sensing module can be used to recognize human motions; i.e., we could make it discriminate the falling-down state from other normal human motions. When falling is recognized, a trigger signal will be sent to a valve to open airbags and the gas is released to inflate the airbags that reduces the force of impact when a person falls down to the ground.

For independent motion recognition, the motion sensors work together with the MCU, as shown in Figure 8.10. The MCU first converts the analog accelerations and angular rates into digital signals. Then, using special algorithms, including coordinate transformation and filters, the system classifies the motion. For initial experiments, we transmitted all motion sensor data to a computer. After analysis, a reliable filter and recognition algorithm was developed and implemented entirely in the MCU, which removes the need for an external computer.

We put μIMU on the left hip of a human and conducted two groups of experiments. The first is a 100 lateral falls since this is the most typical situation that is likely to cause a hip fracture. The second is one 100 other motions, including 10 running, 20 walking, 20 sitting, 20 squating, 10 stepping on stairs, 10 walking in an elevator, and 10 jumping actions. The reason for selecting these motions is that they are commonly encountered motions. For the elderly, who seldom jump and run, we collected more sitting and squatting motion data. In addition, these motions are more similar to the motion during a fall. These experiments could

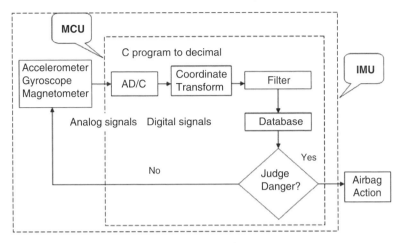

Figure 8.10. Schematic chart of the μIMU system.

then formulate a database for later SVM training and judgment. We can also discriminate falling from other motions directly from analysis of the original data.

A lateral fall of a human subject can be modeled as

$$\frac{1}{6}mL^2\omega^2 + \frac{1}{2}mgL\cos(\varphi) = \frac{1}{2}mgL \tag{8.1}$$

$$\omega^2 = \frac{3g}{l}(1 - \cos\varphi) \tag{8.2}$$

where l is the height of one's hip, ω is the angular rate when a person is falling, and φ is the tilt angle when falling. A static fall is one for which the only energy supplied is gravity. During a fall, rotation is also observed and the angular rate is related to the angle that a person is tilting and the height of the center of mass of the person (Equation 8.2). In static lateral fall experiments that we conducted, the person was assumed to fall from a start where the angular rate is zero.

Figure 8.11 shows fall data recorded by the sensor module. One acceleration and two angular rates are not shown since for an idealized lateral falling motion, the only motion of the body is due to gravity. That is, the acceleration Y (A_y) and acceleration Z (A_z) all change by 1-G during a fall. The angular rate in the lateral direction Gy will also change. We modeled the angular rate change during a fall using Equation 8.2, and the analytical results closely matched the experimental results (Gy) as shown in Figure 8.12. From the figure we can also see that accelerations due to gravity change only at impact and not at the beginning of a fall. This is because the person is first under a free-fall, and the body is effectively weightless during this period. Therefore, there is no change in the sensor signal before the body angle is larger than some value.

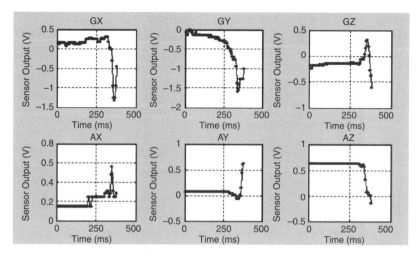

Figure 8.11. Pure falling-down state cutting.

According to this observation, we programmed our microcontroller to compare the angular rate with a fixed threshold. When the angular rate value is larger than this value, we send an activation signal to the airbag for inflation.

We set up a demonstration by connecting the μIMU and the airbag together. When a person is falling, the μIMU transmits a danger signal to the relay, and the relay drives the solenoid valve open and the compressed air will inflate the airbag, reducing the impact force of the fall. Figure 8.13 shows a successful demonstration of the human airbag system. The airbag was inflated just before the person falls to the ground.

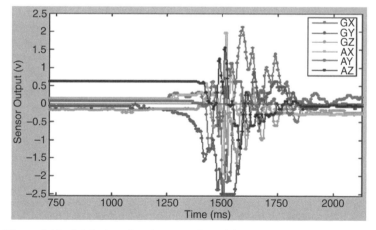

Figure 8.12. Original motion data recording of falling-down. See color insert.

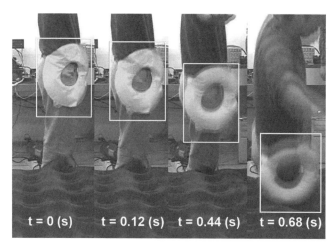

Figure 8.13. A demonstration of real-time recognition of falling-down motion with inflation of airbag. See color insert.

8.3 SUPPORT VECTOR MACHINE FOR HUMAN MOTION DETERMINATION

Although a simple angular rate threshold gives good results, false inflations can occur during normal physical activity. An SVM-based scheme using a host computer was tested to better distinguish between normal and falling motions.

As shown in Figure 8.14, the MCU first converts the sensor outputs to digital signals and then transmits the packed data signal sequentially via a Bluetooth module to a computer. Hundreds of recordings that included lateral falls, walking, running, sitting, walking up and down stairs, and walking in elevators were made to form a database for SVM training. After training, we selected the best features to form a classifier for falling-motion recognition.

Our goal is to recognize falling-down motion in real time in order to control the hip-protection airbag. We address this classification problem as binary pattern recognition with SVMs:

1. Set up a motion database of "falling-down" and "non-falling-down" examples using the μIMU;
2. Use supervised PCA (principle component analysis) to generate and select characteristic features;
3. SVM training (SVM) to produce a classifier.

8.3.1 Principal Component Analysis for Feature Generation

Feature generation and selection are very important in falling-down recognition, as badly selected features such as weightlessness, learning backward, and hip spinning

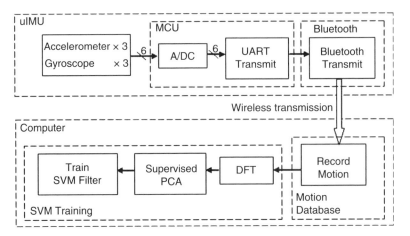

Figure 8.14. Schematic chart of SVM training.

may cause confusion with jumping, sitting, and turning around, which can obviously diminish the performance of the system. Furthermore, even though the current features contain enough information about the output class, they may not predict the output correctly because the dimension of feature space may be so large that it requires numerous instances to determine the result.

PCA can generate mutually uncorrelated features while packing most of the information into several eigenvectors. In our system, we use supervised PCA algorithms to generate features and select high-quality combinations for better recognition performance.

A set of eigenvectors can be computed from the training motion data, and some eigenvectors are selected for classification according to the corresponding eigenvalue.

We selected the eigenvectors according to the binary classified capability, instead of the corresponding eigenvalues. It is because the eigenvectors with large eigenvalues may carry the common features, but not the distinguishing information between the two classes.

The method can be described as follows. Suppose that we have two sets of training samples: A and B. The number of training samples in each set is N. Φ_i represents each eigenvector produced by PCA. Each training sample, including positive samples and negative samples, can be projected into an axis extended by the corresponding eigenvector. By analyzing the distribution of the projected $2N$ points, we can roughly select the eigenvectors that have more motion information. The following is a detailed description of the process:

1. For a certain eigenvector Φ_i, compute its mapping result according to the two sets of training samples. The result can be described as $\lambda_{i,j}$, ($1iM$, $1j2N$).
2. Train a classifier f_i using a simple method such as a perception or neural network, which can separate $\lambda_{i,j}$ into two groups: falling-down and non-falling-down with a minimum error $E(f_i)$.

3. If $E(fi) < \theta$, then we delete this eigenvector from the original set of eigenvectors. M is the number of eigenvectors, and $2N$ is the total number of training samples. θ is the defined threshold. The left few eigenvectors are selected. The eigenvectors can also represented back to motion data representing a typical movement.

The key point is shown in Figure 8.15 for evaluation of eigenvectors. The performance of Eigenvector i is better than that of Eigenvectors 1 and 2.

It is possible that we select too few good eigenvectors, even none in a single PCA analysis process. We propose the following approach to solve this problem. We assume that the number of training samples, $2N$ is large enough. We randomly select training samples from the two sets. The number of selected training samples in each set is less than $N/2$. Then, we perform the supervised PCA analysis with them. By repeating the previous process, we can collect several good features. This approach is inspired by the bootstrap method. The main idea of this approach is that it may emphasize some good features by reassembling data and then make the features stand out easily.

8.3.2 Support Vector Machine Classifier

The SVM is a new technique in the field of statistical learning theory [9]. Originally, SVM was developed from classification problems. It was then extended to regression estimation problems, i.e., to problems related to finding the function: $y = f(\overline{x}), y \in R, \overline{x} \in R^N$, given by its measurements yi with noise at some (usually random) vector $\overline{x}_i, (y_1, \overline{x}_1), \ldots, (y_l, \overline{x}_l)$.

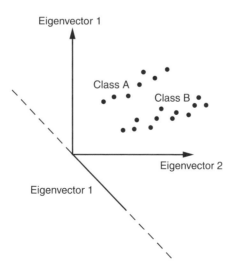

Figure 8.15. The different distinguishing ability of different eigenvectors.

In SVM, the basic idea is to map the data X into a high-dimensional feature space f via a nonlinear mapping Φ, and to do linear regression in this space [10].

$$f(\bar{x}) = (\omega \cdot \Phi(\bar{x})) + b(\Phi : R^N \rightarrow F, \omega \in F) \tag{8.3}$$

where b is a threshold. Thus, linear regression in a high-dimensional (feature) space corresponds to nonlinear regression in the low-dimensional input space R^N. Note that the dot product in Equation 8.3 between ω and $\Phi(\bar{x})$ would have to be computed in this high-dimensional space (which is usually intractable), if we are not able to use the kernel that eventually leaves us with dot products that can be implicitly expressed in the low-dimensional input space R^N. Since Φ is fixed, we determine ω from the data by minimizing the sum of the empirical risk $R_{emp}[f]$ and a complexity term $\|\omega\|^2$, which enforces flatness in feature space:

$$R_{reg}[f] = R_{emp}[f] + \lambda\|\omega\|^2 = \sum_{i=1}^{l} C(f(\bar{x}_i) - y_i) + \lambda\|\omega\|^2 \tag{8.4}$$

where l denotes the sample size $(\bar{x}_1, \ldots, \bar{x}_l)$, $C(.)$ is a loss function, and λ is a regularization constant. For a large set of loss function, Equation 8.4 can be minimized by solving a quadratic programming problem, which is uniquely solvable [11]. It can be shown that the vector ω can be written in terms of the data points:

$$\omega = \sum_{i=1}^{l} (\alpha_i - \alpha_i^*)\Phi(\bar{x}_i) \tag{8.5}$$

with α_i, α_i^* being the solution of the aforementioned quadratic programming problem [10]. α_i and α_i^* have an intuitive interpretation as forces pushing and pulling the estimate $f(\bar{x}_i)$ toward the measurements y_i [11]. Taking Equations 8.5 and 8.3 into account, we can rewrite the whole problem in terms of dot products in the low-dimensional input space

$$f(\bar{x}) = \sum_{i=1}^{l} (\alpha_i - \alpha_i^*)(\Phi(\bar{x}_i) \cdot \Phi(\bar{x})) + b = \sum_{i=1}^{l} (\alpha_i - \alpha_i^*)K(\bar{x}_i, \bar{x}) + b \tag{8.6}$$

where α_i, α_i^* are Lagrangian multipliers and $\bar{x}_i$ are support vectors.

In Equation 8.6, we introduce a kernel function $K(\bar{x}_i, \bar{x}) = \Phi(\bar{x}_i) \cdot \Phi(\bar{x}_j)$. As explained in Ref. 8, any symmetric kernel function K satisfying Mercer's condition corresponds to a dot product in some feature space.

For a detailed reference on the theory and computation of SVM, readers can refer to Ref. 9.

Many kernels satisfy the Mercer's condition as described in Ref. 9. In this project, we take a simple polynomial kernel in Equation 8.6:

$$K(\bar{x}_i, \bar{x}) = ((\bar{x}_i \cdot \bar{x}) + 1)^d \qquad (8.7)$$

where d is user defined (taken from Ref. 7).

After the off-line training process, we obtain the values for Lagrangian multipliers and support vectors of SVM. Let $\bar{x} = [x_1, x_2, \ldots, x_N]^T$ ($\bar{x}_i$ is an element of $\bar{x}$ and is a sample data $\bar{x}_i$ of $\bar{x}$). By expanding Equation 8.6 according to Equation 8.7, we know that $f(\bar{x})$ is a nonhomogeneous form of degree d in $\bar{x} \in R^N$

$$f(\bar{x}) = \sum_{0 \leq i_1 + i_2 + \cdots + i_N \leq d} c_j x_1^{i_1} x_2^{i_2} \cdots x_N^{i_N} \qquad (8.8)$$

where $i_1, i_2, , i_N$ are nonnegative integers and $c_j \in R$ are weighting coefficients. j can be 1,2,...,M, where $M = \begin{pmatrix} N+d \\ N \end{pmatrix}$.

8.4 EXPERIMENTAL RESULTS

Experiments were performed to demonstrate the motion detection in 3D space of our μIMU. We first performed experiments of lateral falling-down and other motions to form the database. With the best features supervised PCA and motion database, we obtained SVM filter after training. The recognition rate can be as high as 100%.

8.4.1 Motion Detection Experiments and Database Forming

Two groups of experiments were performed: 100 times of lateral falling-down and 100 times of other motions, including 10 times running, 20 times walking, 20 times sitting, 20 times squat, 20 times stepping stairs, and 10 times jumping. The reason for selecting these motions is that they are the normal motions of life. For the elderly, they seldom jump and run; hence, we collected more sitting and squatting motion data, in addition to the fact that these motions are more similar to the motion during a fall.

Time-sequenced pictures at an experimental subject during a fall are shown in Figure 8.16. Figure 8.17 shows the original data of motion, including 3D accelerations and rotation rates from one experimental trial. Gy in Figure 8.18 is the angular rate of the pitch direction; hence, we can judge a fall-motion stating from the changing of Gy. Az is acceleration in the vertical direction. A sudden spike in this data corresponds to when the hip hits the ground. By synchronizing visual observations with the sensed data, we extracted the motion data from beginning of a fall to when the body hits the ground (soft mat) for all six sensors. Figure 8.18 is the discrete fourier transform (DFT) result, which records information in frequency domain. One

Figure 8.16. Lateral falling-down state. See color insert.

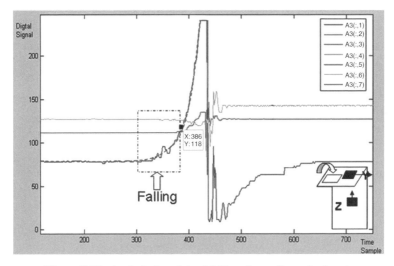

Figure 8.17. Analysis for a lateral falling-down. See color insert.

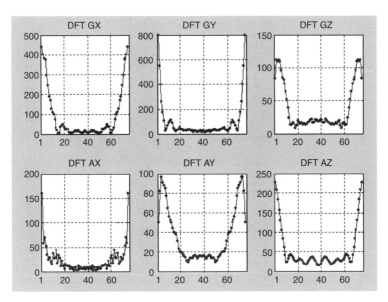

Figure 8.18. DFT transform result VS frequency of failling.

hundred fall-down experiments were performed, including different falling-motions, and the experiments were also performed by two different people for construction of a more realistic database. These experimental data were used to falling-down reference.

Figures 8.19, 8.20 and 8.21 show the corresponding motion data from the sensors when the subject is running. In Figure 8.20 and 8.21, only one cycle of data is extracted for analysis.

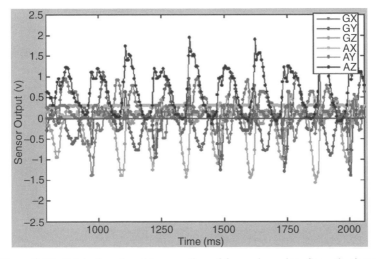

Figure 8.19. Original motion data recording of forward running. See color insert.

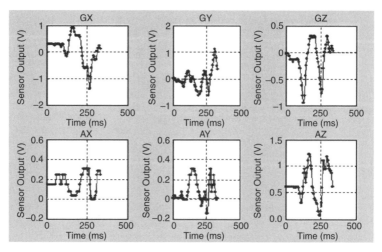

Figure 8.20. One period running state cutting.

8.4.2 SVM Training and Falling-Down Recognition

We recorded 200 experiment results with half "falling-down" and half "non-falling-down." Each result consisted of six arrays measured by the six sensors, respectively.

We performed data preprocessing to filter noise and reduce dimension. For each experimental result, we performed DFT with the six arrays, respectively. We kept

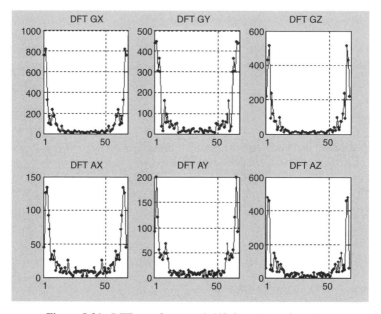

Figure 8.21. DFT transform result VS frequency of running.

Table 8.2. The Coefficient of the SVM Classifier

c_0	-2.084349241	c_5	0.00000498
c_1	0.007279936	c_6	-0.000000173
c_2	-0.003962637	c_7	0.000000374
c_3	-0.002182333	c_8	-0.000001469
c_4	-0.000003207	c_9	0.000002342

the first 10 orders of each DFT result. After 200 times DFT, we obtained a matrix of 200 rows and 60 columns. Each row represents one experiment. Each has 6×10 numbers by the sequence of G_x, G_y, G_z, A_x, A_y, A_z.

Then we selected the good vectors as training features. After 213 supervised PCA, we obtained 175 good eigenvectors. But the feature information by one eigenvector may not be sufficient enough. Thus we randomly selected out three eigenvectors from the 175 eigenvectors and performed the supervised PCA. After 3152 times of selection, we obtained 87 good triple sets. After testaments, we found our measuring and training systems are very good. All top nine eigenvector sets can classify the 200 vectors into "falling-down" and "non-falling-down" with 100% correct.

Then we randomly chose the No. 80 triple set for SVM training and obtained the coefficients of the SVM filter (as shown in Table 8.2).

$$f(x_1, x_2, x_3) = c_0 + c_1 \cdot x_1 + c_2 \cdot x_2 + c_3 \cdot x_3 + c_4 \cdot x_1^2$$
$$+ c_5 \cdot x_1 \cdot x_2 + c_6 \cdot x_1 \cdot x_3 + c_7 \cdot x_2^2 + c_8 \cdot x_2 \cdot x_3 + c_9 \cdot x_3^2 \tag{8.9}$$

The test result shows that this SVM classifier can categorize these 200 experiment results with 100% correct. And because the filter is in the form of algorithm operation, the calculation time is very short.

8.5 CONCLUSION

This chapter presents a novel MEMS-based human airbag system that is under development. A μIMU for detection of complex human motions and recognition of falling-down motion is used, which can then be used to trigger the release of airbags. Experiments showed that our airbag system can achieve real-time recognition and a fast response, which ensures that the airbags can be released before a human impacts the ground. The μIMU can also measure human motion in the form of accelerations and rotations in three dimensions. With a Bluetooth module, the small size unit can transmit experimental data to a computer for postexperimental data analysis. We also used SVM as a pattern recognition method for training of data transformed via a DFT and PCA. We have shown that selected eigenvector sets can classify 200 experimental data sets that can be used to classify the eigenvectors into "non-falling-down" or "falling-down" categories without error.

REFERENCES

1. M. N. Nyan, F. E. H. Tay, T. H. Koh, Y. Y. Sitoh, and K. L. Tan, "Location and sensitivity comparison of MEMS accelerometers in signal identification for ambulatory monitoring," in *Proc. Electronic Components and Technology*, Vol. 1, June 1–4, 2004, pp. 956–960.

2. C. Baudoin, P. Fardellone, K. Bean, et al., "Clinical outcomes and mortality after hip fracture: a 2-year follow-up study," *Bone*, Vol. 18, Suppl. 3, 1996, pp. S149–S157.

3. A. H. F. Lam, W. J. Li, Y. Liu, and N. Xi, "MIDS: micro input devices system using MEMS sensors," in *Proc. IEEE/RSJ International Conference on Intelligent Robots and System 2002*, Vol. 2, Sept. 30 - Oct. 5, 2002, pp. 1184–1189.

4. A. H. F. Lam, R.H.W. Lam, W. J. Li, M. Y. Y. Leung, and Y. Liu, "Motion sensing for robot hands using MIDS," in *Proc. IEEE International Conference on Robotics and Automation*, Vol. 3, Sept. 14–19, 2003, pp. 3181–3186.

5. J. Mantyjarvi, J. Himberg, and T. Seppanen, "Recognizing human motion with multiple acceleration sencors," in *Proc. of IEEE Conf. on Systems Man. and Cybernetics*, Oct. 2001, pp. 747–752.

6. ATMEL Corporation, 8-bit AVR mircrocontroller with 32K bytes in-system programmable flash datasheet, 2004.

7. Analog Devices Inc. Precision $\pm 17\, g$ single/dual axis accelerometer ADXL203 datasheet, 2004.

8. Analog Devices Inc. $\pm 300^\circ/s$ single chip yaw rate gyro with signal conditioning ADXR300 datasheet, 2004.

9. S. Bernhard, C. J. C. Burges, and A. Smola, *Advances in Kernel Methods Support Vector Learning*, Cambridge, MA, MIT Press, 1998.

10. N. Cristianini and J. Shawe-Taylor, *An Introduction to Support Vector Machines and Other Kernel-Based Learning Methods*, Cambridge University Press, Cambridge, 2000.

11. A. Smola, "General cost function for support vector regression," in *Proc. of the Ninth Australian Conf. on Neural Networks*, 1998, pp. 79–83.